NEUROSCIENCE RESEARCH PROGRESS

MOTOR BEHAVIOR AND CONTROL

NEW RESEARCH

NEUROSCIENCE RESEARCH PROGRESS

Additional books in this series can be found on Nova's website
under the Series tab.

Additional e-books in this series can be found on Nova's website
under the e-book tab.

NEUROSCIENCE RESEARCH PROGRESS

MOTOR BEHAVIOR AND CONTROL

NEW RESEARCH

MARCO LEITNER

AND

MANUEL FUCHS

EDITORS

New York

For permission to use material from this book please contact us:
Telephone 631-231-7269; Fax 631-231-8175
Web Site: http://www.novapublishers.com

Library of Congress Cataloging-in-Publication Data

ISBN: 978-1-62808-142-8

Library of Congress Control Number: 2013940383

Published by Nova Science Publishers, Inc. † New York

Contents

Preface

In this book the authors present new research in the study of motor behavior and control. Topics discussed include motor planning skills in neurotypical development and Autism Spectrum Disorder; the role of motor imagery in action planning; cognitive training as how it enhances motor performance and learning; laterality, load and motor imagery; motor control and impulsivity; the contribution of better understanding of martial arts strikes to studies in motor control; assisted cycle therapy (ACT); developing the ability to activate difficult facial action units involved in emotional expressions; and coordination thermodynamics.

Chapter 1 - As an almost infinite number of options are available to complete any given task, of particular interest is how a person is constrained to perform a movement in a specific manner. This exemplifies the degrees-of-freedom problem. That said, the movement-selection system must also take into consideration a motor plan, which facilitates the most efficient movement. Therefore, movement selection is influenced by movements completed in the recent past, where efficient movements are implemented to minimize the costs associated. End-state comfort is defined as the tendency to maximize comfortable hand and arm postures at the end of simple object manipulation tasks rather than at the start; therefore, the end-state comfort effect is used to assess behavioural evidence of advanced motor planning. More specifically, how a movement is planned prior to initiation. Within the developmental literature, it is suggested that end-state comfort is not likely an innate characteristic, but a motor phenomenon acquired with age. Neurotypical children thus perform tasks in a manner which highlights their lack of sensitivity to end-state comfort. Recent investigations suggest that the ability to differentiate between comfortable and uncomfortable grips is fully mature by the age of 9, where adult-like sensitivity to end-state comfort is also observed. The end-state comfort effect can also be applied to a joint action paradigm, in order to better understand how one person anticipates the motor intentions of another. Adults have been observed planning joint action movements based on their coactors' needs. More specifically, manipulating one's own end-state comfort to ensure an object is passed in a manner that facilitates beginning-state comfort of another. With respect to children, adult-like patterns of beginning-state comfort have been observed by the age of 7. Therefore, it can be suggested that 7-year-olds consider another's beginning-state comfort; however, are unable to facilitate their own end-state comfort until age 9.

Neurotypical development can be used as a means of comparison to developmental impairment. For example, it is suggested that motor planning deficits exist in Autism Spectrum Disorder (ASD). A number of researchers have observed delayed movement initiation and movement execution times in adolescents with ASD, while performing various high accuracy pointing and choice reaction time tasks. Additionally, Fabbri-Destro and colleagues observed an impaired ability of adults with ASD to chain together movement sequences. Specifically, they examined that the initial stage of a multi-step movement sequence was not modulated by the complexity of the end goal in this population, contrary to the behaviour of healthy adult controls. With respect to end-state comfort specifically, Hughes showed that adolescent participants with ASD exhibited a predictive movement impairment when asked to pick up a bar and orient it a certain way into a metal disk. The author attributed this to a deficit in predictive movement or motor planning impairment. Research in the authors' lab has demonstrated that motor planning deficits in ASD exist, as children with ASD displayed significantly fewer instances of end-state comfort and beginning-state comfort than their neurotypical counterparts; however, trends in development and subsequent manifestations in young adulthood have yet to be delineated.

The purpose of this chapter is thus two-tiered. First, the authors aim to discuss the underlying factors influencing the neurotypical development of end- and beginning-state comfort from early childhood to young adulthood. It is then of additional interest to discuss the developmental time course in ASD in comparison to neurotypical counterparts.

Chapter 2 – Motor imagery is a widely used experimental paradigm for the study of cognitive aspects of action planning and control in adults. Underscoring that interest is the assertion that motor imagery provides a window into the process of action representation. These notions complement internal modeling theory suggesting that such representations allow predictions (estimates) about the mapping of the self to parameters of the external world; processes that enable successful planning and execution of action. Those observations have drawn the attraction of developmentalists that work with typically developing children and special populations. This chapter defines motor imagery and its link to mental representation, internal modeling, embodied cognition, and foremost, action planning. Included is a selection of recent work with typically developing children and special populations. The merits of this review are associated with the apparent increasing attraction for studying and using motor imagery to understand the developmental aspects of action processing in children.

Chapter 3 – Cognition and motor performance are interdependent for optimal daily functioning. In recent years, the role of cognitive training for able-bodied and disabled populations has been widely investigated. The importance of cognitive training for cognitive and motor improvement has generally received strong empirical support. There is evidence that non-trained domains can also be enhanced (e.g., sport-specific creativity and driving). The findings suggest that motor performance and learning can be improved when viable cognitive training paradigms are used. In this chapter, we first examine the roles of attention, memory, and imagery on motor performance and learning. Second, we discuss how cognitive training can enhance motor abilities. Third, the feasibility and prospect of computerized cognitive training on motor performance and learning is considered. Finally, practical implications related to cognitive training, motor performance, and rehabilitation are examined as related to the need for future research.

Chapter 4 – Actual and imagined movement durations (MDs) are equivalent when inertial loads are low relative to the maximum inertial loading capacity of (dominant) index finger movement, but as load increases imagined MDs lengthen at a faster rate than actual MDs. The load-dependent lengthening of imagined over actual MD may arise from increased uncertainty in predicting motor output: Uncertainty in predicting movement outcomes has been correlated with increases in MD. Here, the non-dominant and dominant index fingers participated in actual and imagined action under different loads. Because the non-dominant finger may have less experience moving heavier loads—with increased uncertainty in controlling those loads—it was hypothesized that the actual-imagined MD gap should increase at a faster rate for the non-dominant finger. The results for both fingers replicated prior research. However, there were no laterality effects. Movement in the current task was limited to a single biomechanical degree of freedom. A possible explanation for the absence of a laterality effect is that the non-dominant and dominant controllers are equally effective in controlling systems with few elements.

Chapter 5 – Present investigations have demonstrated Motor Control as a possible field for evaluating the adaptive properties of impulsivity. In this chapter the authors present: (a) some possible definitions of impulsivity, (b) the negative impact of impulsivity on motor control, (c) evidences indicating the existence of functional properties of impulsivity, and (d) the first findings indicating an adaptive, functional role of impulsivity on motor control.

Although generally viewed as counterproductive behavior, it seems that impulsivity has a positive impact on motor control in some specific circumstances, in a fast-paced context that requires quick decision-making and fast movements, implicit/automatic processing observed in high-impulsive subjects seems to be productive.

Overall, the relationship between impulsivity and motor control depends on the sensory-motor aspects of the task.

Chapter 6 – Martial Artists distinguish themselves from regular people as they constantly train their bodies to be strong, precise, accurate and fast. However, an increase in one of these characteristics may be detrimental to the others. Additionally, quantifying striking peak force is most often challenging and determining what other physical variables are reliable when trying to understand within-subject and between-subject peak force variations is worthwhile. As such, this study has two main goals. The first goal is to investigate the possible trade off between peak hand acceleration and accuracy and consistency of hand strikes performed by martial artists of different training experiences. The second is to investigate the correlation among several physical variables commonly used to quantify the performance of martial arts and combative sports strikes. Thirteen martial artists (10 male and 3 female) volunteered to participate in the experiment. Each participant performed 12 maximum effort goal-directed strikes targeted at an instrumented pendulum. The target was instrumented with one load cell, a pressure sensor, and a tri-axial accelerometer block. Additionally, hand acceleration during the strikes was obtained using tri-axial accelerometer block. The authors estimated the subject's accuracy, precision, hand's speeds before impact, peak accelerations before and during impact, the strike's peak force, the pendulum's peak acceleration. They found that for our male subjects training experience was significantly correlated to hand peak acceleration prior to impact (r^2 = 0.456, p = 0.032) and accuracy (r^2 = 0. 621, p = 0.012). These correlations suggest that more experienced participants exhibited higher hand peak accelerations and at the same time were more accurate. Training experience, however, was not correlated to consistency (r^2 = 0.085, p = 0.413). Overall, these results suggest that martial arts training may lead practitioners to achieve higher striking hand accelerations with better accuracy and no change in striking consistency. Furthermore, considering within-subject variations, the authors found that peak hand acceleration during the impact (mean r^2=0,56) followed by peak hand speed before impact (r^2=0.33) were the variables that exhibited greater correlation to peak force. As for between-subject comparisons, they found that peak hand acceleration before impact (r^2=0.82) and hand speed before impact (r^2=0.82) were highly correlated to peak force.

Chapter 7 – Assisted Cycle Therapy (ACT) is an innovative exercise in which the participant pedals on a bicycle at 35% greater than their preferred cycling rate with the assistance of a mechanical motor. Previous research in Parkinson's Disease patients found improvements in bimanual dexterity (e.g., grasping forces, interlimb coordination) and clinical measures of movement (e.g., UPDRS) after ACT but not after voluntary exercise or no exercise. Recent research with adolescents with Down syndrome found improvements in manual dexterity as measured by the Purdue Pegboard after an acute 30 minute bout of ACT but not after similar Voluntary or No exercise sessions. Improvements in the upper extremity functioning when the lower extremities were exercised suggests that changes are occurring at the cortical level to create improvements in global motor control. Possible central mechanisms include neurogenesis caused by upregulation of neurotrophic factors (e.g., BDNF) or increased sensory input to the motor cortex due to the high pedaling rate. Neurologic disorders that inhibit movement rate are suggested to benefit from ACT. The implications for improving motor, cognitive, clinical and health outcomes in several neurologic disorders will be discussed.

Chapter 8 – Although adults are able to activate most of the action units involved in emotional expressions voluntarily, there are some action units that have proven to be very difficult to activate. In this paper, the authors investigated the effect of practice on the voluntary control of three action units: cheek raiser, upper lip raiser, and lip corner depressor. Twenty young adults were given 25 training trials to activate these action units, and their performance was assessed with the Facial Action Coding System. The results indicate an effect of practice for the upper lip raiser only. As predicted, several non-target action units were activated when the participants performed the task, and they were consistent with those found in past research.

Chapter 9 – Human motor behavior frequently requires a large degree of coordination. Catching requires motor coordination with an environmental stimulus. Walking and stair climbing requires inter-limb coordination. Motor activities in groups of people such as team sports activities require between-persons coordination of movements. The deterministic laws of human motor coordination have been successfully described in terms of dynamical systems. In contrast, the formulation of thermodynamic laws of motor coordination is still a challenge for modern day science. The reason for this is that a general thermodynamic theory for non-equilibrium systems such as human motor control systems is not available. In order to address this problem, it has recently been suggested to consider stochastic descriptions of human motor coordination that are grounded in linear non-equilibriumthermodynamics. Two special cases are the socalled canonical-dissipative approach to uni-manual motor control problems and the mean field theoretical approach to motor coordination in groups.

The general theoretical framework of a theory of coordination thermodynamics is outlined. In addition, the coordination thermodynamical perspective is illustrated for two experimental studies on single person motor coordination and between-persons motor coordination of groups.

In: Motor Behavior and Control: New Research
Editors: Marco Leitner and Manuel Fuchs
ISBN: 978-1-62808-142-8

Chapter 1

Motor Planning Skills in Neurotypical Development and Autism Spectrum Disorder: End- and Beginning-State Comfort

Sara M. Scharoun,[1*] *Jessica L. Pinkerton*[2] *and Pamela J. Bryden*[2]
[1]Department of Kinesiology, University of Waterloo, Ontario, Canada
[2]Department of Kinesiology and Physical Education, Wilfrid Laurier University, Ontario, Canada

Abstract

As an almost infinite number of options are available to complete any given task, of particular interest is how a person is constrained to perform a movement in a specific manner. This exemplifies the degrees-of-freedom problem (Bernstein, 1967). That said, the movement-selection system must also take into consideration a motor plan, which facilitates the most efficient movement (Jordan and Rosenbaum, 1989). Therefore, movement selection is influenced by movements completed in the recent past, where efficient movements are implemented to minimize the costs associated (Rosenbaum and Jorgensen, 1992; Fischer, Rosenbaum and Vaughan, 1997). End-state comfort is defined as the tendency to maximize comfortable hand and arm postures at the end of simple object manipulation tasks rather than at the start (Rosenbaum and Jorgensen, 1992; Rosenbaum, Vaughan, Barnes and Jorgensen, 1992); therefore, the end-state comfort effect is used to assess behavioural evidence of advanced motor planning. More specifically, how a movement is planned prior to initiation (Rosenbaum, van Heugten and Caldwell, 1996). Within the developmental literature, it is suggested that end-state comfort is not likely an innate characteristic, but a motor phenomenon acquired with age (e.g. Adalbjornsson, Fischman and Rudisil, 2008). Neurotypical children thus perform tasks in a manner which highlights their lack of sensitivity to end-state comfort. Recent

* Corresponding Author: Sara M. Scharoun sscharou@uwaterloo.ca.

investigations suggest that the ability to differentiate between comfortable and uncomfortable grips is fully mature by the age of 9, where adult-like sensitivity to end-state comfort is also observed (Scharoun and Bryden, in press; Stöckel, Hughes and Schack, 2011). The end-state comfort effect can also be applied to a joint action paradigm, in order to better understand how one person anticipates the motor intentions of another. Adults have been observed planning joint action movements based on their coactors' needs (Gonzalez, Studenka, Glazebrook and Lyons, 2011; Ray and Welsh, 2011). More specifically, manipulating one's own end-state comfort to ensure an object is passed in a manner that facilitates beginning-state comfort of another. With respect to children, adult-like patterns of beginning-state comfort have been observed by the age of 7 (Scharoun and Bryden, in press). Therefore, it can be suggested that 7-year-olds consider another's beginning-state comfort; however, are unable to facilitate their own end-state comfort until age 9.

Neurotypical development can be used as a means of comparison to developmental impairment. For example, it is suggested that motor planning deficits exist in Autism Spectrum Disorder (ASD). A number of researchers have observed delayed movement initiation and movement execution times in adolescents with ASD, while performing various high accuracy pointing and choice reaction time tasks (e.g. Rinehart, Bradshaw, Brereton and Tonge, 2001; Glazebrook, Elliot, and Lyons, 2006; Rinehart et al., 2006; Dowd, McGinley, Taffe and Rinehart. 2012). Additionally, Fabbri-Destro and colleagues (2009) observed an impaired ability of adults with ASD to chain together movement sequences. Specifically, they examined that the initial stage of a multi-step movement sequence was not modulated by the complexity of the end goal in this population, contrary to the behaviour of healthy adult controls (Fabbri-Destro, Cattaneo, Boria and Rizzolatti, 2009). With respect to end-state comfort specifically, Hughes (1996) showed that adolescent participants with ASD exhibited a predictive movement impairment when asked to pick up a bar and orient it a certain way into a metal disk. The author attributed this to a deficit in predictive movement or motor planning impairment (Hughes, 1996). Research in our lab has demonstrated that motor planning deficits in ASD exist, as children with ASD displayed significantly fewer instances of end-state comfort and beginning-state comfort than their neurotypical counterparts (Scharoun and Bryden, 2013); however, trends in development and subsequent manifestations in young adulthood have yet to be delineated.

The purpose of this chapter is thus two tiered. First, we aim to discuss the underlying factors influencing the neurotypical development of end- and beginning-state comfort from early childhood to young adulthood. It is then of additional interest to discuss the developmental time course in ASD in comparison to neurotypical counterparts.

Introduction

A seemingly simple behaviour, such as enjoying a freshly brewed morning coffee may seem effortless, almost second nature. However, goal-directed movements involve the coordination of joint and body segments based on a pre-determined sequence of movement. Motor planning has been described as, "the process of converting a current state (my hand is by my side) and desired state (my hand should be on the mug) into a sequence of motor commands (move the arm, close the fingers)" (Gowen and Hamilton, 2013, p. 333). As an almost infinite number of movement options are available to complete any given task, of particular interest is how an individual is constrained to perform such movement in a specific manner. This exemplifies the degrees-of-freedom problem (Bernstein, 1967).

In order to complete a movement, the movement-selection system must consider what is required to facilitate the most efficient movement (Jordan and Rosenbaum, 1989). Therefore, movement selection is influenced by movements completed in the recent past (Kent, Wilson, Plumb, Williams, and Mon-Williams, 2009) where efficient movements are implemented to minimize the costs associated (Rosenbaum and Jorgensen, 1992; Fischer, Rosenbaum and Vaughan, 1997). In line with this, it is suggested that a desire to move with minimal rotation (Cruse, 1986; Cruse, Wischmeyer, Brüwer, Brockfeld, and Dress, 1990; van Bergen, van Swieten, Williams, and Mon-Williams, 2007) constrains movement. To complete an efficient movement, "the hand is rotated through the minimum distance required to reach a final posture" (van Swieten et al., 2010, p. 494).

In addition, end-state comfort must be taken into consideration (Rosenbaum et al., 1990). Rosenbaum and colleagues (1990) used a bar transport task to exemplify the end-state comfort effect. The task consisted of a 'bar' painted half black and half white, resting horizontally on a 'cradle' placed on a table. Participants "could easily lift the bar from the cradle, using either an overhand grip (approaching the bar from above) or an underhand grip (approaching the bar from below)" (Rosenbaum and Jorgensen, 1992, p. 63). The four tasks required participants to "place the left or the right end of the bar squarely on the disk to the left or right of the cradle" (Rosenbaum and Jorgensen, 1992, p. 63). When performing the bar transport task, which required pronation and supination, adults preferred to end their movement with a neutral posture (an overhand grip; thumb down). Therefore, the initial position of the thumb is associated with the rotation of the bar, where selecting an uncomfortable initial posture (underhand grip; thumb up) enables the hand to be returned to a comfortable posture (overhand grip; thumb down) upon completion of the movement (Rosenbaum et al., 1990).

The end-state comfort effect has since been used to assess behavioural evidence of advanced motor planning (e.g. Cohen and Rosenbaum, 2004; Haggard, 1998; Rosenbaum, van Heugten and Caldwell, 1996; Rosenbaum and Jorgensen, 1992; Rosenbaum, Vaughan, Barnes and Jorgensen, 1992; Short and Cauraugh, 1996; Weigelt, Kunde, and Prinz, 2006) by observing how an individual plans a movement prior to completing that movement. Both the Knowledge Model (Rosenbaum, Engelbrecht, Bushe, and Loukopolous, 1993) and the Posture-Based Model (Rosenbaum, Loukopolous, Meulenbroek, Vaughan, and Engelbrecht, 1995; Rosenbaum, Melenbroek, Vaughan, and Jansen, 2001) provide an explanation for the end-state comfort effect. The Knowledge Model takes into consideration storage, planning, and execution—the actor identifies a target, plans and executes the movement all based on prior knowledge of the movement (see Rosenbaum et al., 1993 for a detailed explanation). In comparison, the Posture-Based Model explains that movements are planned based on the final goal of the action as long as the actor is free to select the appropriate motor plan (see Rosenbaum et al., 2001, 1995 for a detailed explanation).

Recent investigations have expanded upon the end-state comfort effect with application to joint action paradigms in order to better understand how one person anticipates the motor intentions of another. For example, when passing a cup of coffee to a colleague, one must consider the intentions of the recipient. Will he/she drink the coffee, or perhaps add cream and sugar? Awareness for this information may ultimately lead to the coffee mug being passed in a manner which facilitates, or hinders, the intended action of the recipient (Gonzalez, Studenka, Glazebrook and Lyons, 2011). Research has demonstrated that adults will ensure an object is passed in a manner that facilitates beginning-state comfort for the recipient, despite the associated cost. Therefore, joint action movements are planned according to coactors' needs and often override the underlying the needs of the actor (Gonzalez et al., 2011; Ray and Welsh, 2011).

The purpose of this chapter is two tiered. First, we aim to discuss neurotypical development of end- and beginning-state comfort. Secondly, as motor planning skills are thought to be impaired in numerous neurodevelopmental disorders, we will discuss the development of motor planning in children with Autism Spectrum Disorders in comparison to their neurotypical counterparts.

End-State Comfort in Neurotypical Development

The bulk of research on end-state comfort has centered around adult participants, thus research from a developmental perspective is limited. Developmental literature to date suggests that end-state comfort is not likely an innate characteristic, but a motor phenomenon acquired with age (Adalbjornsson, Fischman, and Rudisill, 2008; Manoel, and Moreira, 2005; Scharoun and Bryden, in press; Thibaut and Toussaint, 2010; Weigelt and Schack, 2010). Consequently, numerous researchers have observed that young neurotypical children perform motor planning tasks in a manner which emphasizes a lack of sensitivity to end-state comfort. More specifically, different grip selection strategies have been observed, which exemplify both physical and cognitive limitations of advanced planning abilities. For example, Adalbjornsson and colleagues (2008, p. 39-40) observed five grip patterns in an overturned glass task, which did not facilitate end-state comfort: (a) supination strategy (thumb up grasp, using supination to rotate hand away from the body, to turn it over); (b) start-state comfort strategy (thumb up grasp, using supination to rotate hand toward the body to turn it over); (c) chest strategy (thumb up grasp, using the chest for support to turn it over); (d) table strategy (grasp, place on table and re-grasp to turn it over); and (e) top-and twist strategy (grasp cup from top and use finger to twist cup to turn it over).

Motor planning has been explored in infants as young as 9-months of age. McCarty, Clifton, and Collard (1999) presented 9-, 14- and 19-month olds with a spoon full of food at the midline, where handle orientation varied between trials (to the left or right). The ultimate goal—eating—could thus be achieved in a variety of ways, based on the infant's motor plan. In this context, a radial grip with the thumb facing the bowl of the spoon would ultimately result in the most effective movement. However, in 9-month-olds, an ulnar grip was observed 50% of the time, with the thumb facing away from the bowl of the spoon. By 19-months, however, children were able to transport food to the mouth in an efficient/effective manner.

In a follow-up study (McCarty, Clifton, and Collard, 2001), tool use was observed in 9-, 14-, 19- and 24-month-olds during self-directed (e.g. feeding oneself) and other directed (e.g. feeding another) tasks. Results revealed that 14-month-olds displayed an efficient radial grip. However, children were more successful in self-directed conditions, as it is easier to interpret the consequences of such movements (McCarty et al., 2001; Claxton, McCarty, and Keen, 2009). Feeding another (e.g. stuffed animal) did not provide a clear end-state, as the child's perspective of the spoon was the same before and after the goal was achieved. However, in the self-directed task, the food left the spoon and entered the child's mouth, providing feedback—tactile and taste—that the goal was achieved. Therefore, toddlers demonstrated successful motor planning in self-directed actions, as it was easier to interpret the consequences of the movement when a clear end-state was established (McCarty et al., 2001; Claxton et al., 2009).

While the previous studies examined motor planning in toddlers, the breadth of research pertaining to the development of end-state comfort specifically, involves children between the ages of 2- to 12-years-olds. For example, Smyth and Mason (1997) had 4- to 8-year-old children rotate a bar, which had various starting positions, into a target orientation. The authors observed whether or not children adapted their initial hand posture, in order to facilitate a comfortable end posture. Results revealed that planning improves with age; however, adult-like patterns of end-state comfort were not obtained at age 8.

Manoel and Moreira (2005) investigated how right-handed 2.5- to 6-year-old children insert the distal end of a bar into a box of the same shape from a horizontal resting position to a vertical end position. Both low precision (distal end of bar were cylindrical) and high precision (distal end of bar were semi-cylindrical) conditions were implemented. Results revealed that all children displayed a right, overhand grip in both low and high precision conditions. Consequently, it was suggested that hand preference constrains motor planning in young children, in comparison to adults, who typically plan a movement in order to end in a comfortable posture. Therefore, children between the ages of 2.5- and 6-years do not consider end-state comfort, in this context (Manoel and Moreira, 2005).

Adalbjornsson and colleagues (2008) also observed a lack of planning abilities in young children; however, the authors opted for a more natural setting. Using the overturned glass task (Fischman, 1997) with preschool children (2- to 3-year-olds) and kindergarten students (5- to 6-year-olds), results revealed five unique patterns for grip selection: end-state comfort, supination, start-state comfort, chest, table, and top-and-twist. Of the aforementioned strategies, only 11 of 40 children displayed end-state comfort. Overall, both studies (Adalbjornsson et al., 2008; Manoel and Moreira, 2005) suggested that young children demonstrate a lack of planning abilities. However, Manoel and Moreira's (2005) study was limited, such that only six children were in each group. Furthermore, Adalbjornsson and colleagues (2008) implemented an overturned glass task, which required children to skillfully coordinate the two hands when picking up the cup and pouring water; therefore, task complexity may have hindered children's performance.

Weigelt and Schack (2010) used these results as a basis to investigate why children fail to display end-state comfort and, subsequently, how evidence of end-state comfort emerges over the course of sensory-motor development. Paralleling previous studies with adult populations, a bar transport task (Rosenbaum et al., 1990) was used to assess end-state comfort in 3-, 4- and 5-year-olds. In comparison to previous findings, Weigelt and Schack (2010) were the first to show improvement in sensitivity to end-state comfort as a function of age. Furthermore, results highlighted a pattern of improvement for anticipatory planning skills, which are undoubtedly linked to the development of general cognitive control processes, and typically appear between the ages of 5 and 6 (Weigelt and Schack, 2010).

During the same year, Thibaut and Toussaint (2010) assessed 4-, 6-, 8-, 10-year-olds and adults, in order to delineate the age at which adult-like sensitivity to end-state comfort emerges. Bar and pencil manipulation tasks were included, where conditions were manipulated to increase the precision required to complete the movement. Results offered further support to previous suggestions (e.g. Adalbjornsson et al., 2008; Manoel and Moreira, 2005; Weigelt and Schack, 2010) that end-state comfort is a motor phenomenon acquired with age. Furthermore, the authors added to the literature, observing 6-year-olds move faster and with improved accuracy than 8-year-olds in the less constraining task (bar-transport); however, the opposite was true for more constraining task (pencil retrieval to trace an alley). As such, it was suggested that 8-year-old children display motor re-organization patterns, which enable them to incorporate external cues in order to successfully plan a movement. This undoubtedly lays the foundation for adult-like sensitivity to end-state comfort, which was observed in 80 to 90% of 10-year-old children in this context (Thibaut and Toussaint, 2010). Therefore, it can be suggested that sensitivity for end-state comfort develops as a function of sensory-motor development.

Taking the aforementioned into consideration, recent investigations have aimed to further elucidate the development of end-state comfort, while delineating why children fail to demonstrate sensitivity to end-state comfort. Stö ckel, Hughes and Schack (2011) questioned whether planning abilities are associated with the inability to anticipate future events, and/or the failure to differentiate between an initial grasp, which facilitate a comfortable or uncomfortable end posture. Furthermore, whether planning skills are associated with the development of cognitive representations (mental images) underlying the movements. Results revealed that the ability to discriminate between a comfortable and uncomfortable grasp is fully mature in 9-year-olds. This extends that which was previously reported in the literature. Adult-like sensitivity to end-state comfort was also observed at this age, in a bar-transport task. Therefore, cognitive representations of grasp posture unfold throughout development, and are associated with the acquisition of motor planning skills, such as sensitivity to end-state comfort (Stö ckel et al., 2011).

A recent study in our lab (Scharoun and Bryden, in press) complements Stö ckel and colleague's (2011) findings, such that adult-like sensitivity to end-state comfort was observed in 9-year-old participants. Instead of a traditional bar-transport task, we opted to investigate a 'real-world,' naturalistic overturned glass task—similar to Adalbjornsson and colleagues (2008; Fischman, 1997)—but modified to include both an upright glass, and a joint action portion. More specifically, we asked 3- to 12-year-old children and adults to pick-up a cup and (a) pour a glass of water from a pitcher or (b) pass it to the researcher to pour glass of water. Both self- and other-directed movements were investigated, as research has shown children are better able to relate objects to their own body (Lockman and Ashmead, 1983; Rochat, 1998). Paralleling previous observations (e.g. Claxton et al., 2009; McCarty et al., 2001) children's motor planning skills were observed to be more proficient in self-directed actions. This was exaggerated in young children (3- to 8-year-olds; Scharoun and Bryden, in press).

A repeated-measures design was implemented to assess whether children display consistent grip selection tendencies over time, considering inconsistent hand use tendencies (e.g. Bryden, Roy, and Spence, 2007) and various patterns of grip selection (e.g. Adalbjornsson et al., 2008) have been noted in the literature. Our study (Scharoun and Bryden, in press) observed significant improvements in sensitivity to end-state comfort. For example, when manipulating the overturned glass, more end-state comfort was displayed in the second session. Furthermore, when manipulating the cup oriented right-side up, 5- to 6-year-old participants showed more end-state comfort than 3- to 4-year-olds and less than 9- to 12-year-olds and adults in the first session; however, in the second session, 5- to 6-year-olds were not significantly different from older participants. Results thus illustrate that children learn by means of trial-and-error as they explore their environment.

Overall, our recent work (Scharoun and Bryden, in press) in combination with literature to date highlights the link between movement capabilities and cognitive development. Adults display end-state comfort consistently, whereas children show significant improvements in sensitivity within testing sessions as a function of age (Scharoun and Bryden, in press). Therefore, learning to solve the degrees of freedom problem is a key component in the acquisition of motor planning skills (Adalbjornsson et al., 2008). Children enter the world with limited prehensile abilities (Kuhtz-Buschbeck et al, 1998); therefore, grip selection may reflect a bias towards start-state comfort (Manoel and Moreira, 2005; van Swieten et al., 2010). A grip which facilitates start-state comfort does not necessarily reflect planning deficits, but highlights the developmental motor status of the child, such that reaching behaviours parallel experience and ability (van Swieten et al., 2010). Children "use motor skills to explore their environment and it may be through a trial-and error-process that they discover how to solve problems and reach the most efficient motor solutions" (Adalbjornsson et al., 2008, pp. 40). The literature has also revealed that children are better able to facilitate end-state comfort during self-directed movements (Claxton et al., 2009; McCarty et al., 2009; Scharoun and Bryden, in press) in comparison other directed movements. This can be explained by egocentrism—the inability to visualize the perspective of others—(Piaget, 1953; Payne and Issacs, 2012) and/or lack of consideration for negative consequences (e.g. Claxton et al., 2009; McCarty et al., 2001). For example, in an overturned glass task (e.g. Scharoun and Bryden, in press), when asked to pour water, negative consequences highlight to need to plan ahead (Claxton et al., 2009), such that children can (and do) spill water when making last minute adjustments to pour into a cup. That said, a clear end-state is established when water exits the pitcher and enters the cup. In comparison, when asked to pass a cup to the researcher to pour water, the goal movement—the researcher pouring water—is beyond the consideration of young children, therefore, failing to demonstrate end-state comfort has no negative effect on children's movements. However, as children garner the ability to perceive other's intentions (Payne and Isaacs, 2012; Piaget, 1953), more end-state comfort is displayed in other-directed tasks. Overall, "motor planning works as a blind watchmaker, with actions reflecting a previous history of motor evolution where useful actions have survived and less useful ones have perished" (van Swieten et al., 2010, pp. 498). That said, it is argued that grasp selection for certain objects (i.e. a cup) may be a habitual, stimulus-driven action, as opposed to a planned movement (Herbort and Butz, 2011). Therefore it is suggested that observations of overturned-glass tasks (e.g. Adalbjornsoon et al., 2008; Scharoun and Bryden, in press) cannot be generalized to 'neutral' objects used in bar-transport tasks. Knudsen, Henning, Wunsch, Weigelt and Aschersleben (2012) compared the sensitivity to end-state comfort in 3- to 8-year-old children within both bar-transport task, and overturned-glass tasks. The authors observed an increase in sensitivity to end-state comfort from 13% in 3-year-olds to 94% in 8-year-olds within the bar-transport task. However, sensitivity to end-state comfort increased from 63% in 3-year-olds to 100% in 8-year-olds in the overturned-glass task; thus illustrating that the cup may be considered a more familiar object (Knudsen et al., 2012). Children typically begin to use cups around the age of 8-months. However, the required stability to drink from an open cup is not fully mature until a child's third birthday (Carruth and Skinner, 2002). Therefore, the cup may be a familiar object for some 3- to 4-year-olds, whereas, interacting with an open cup may be a novel task for some children in this age group. For example, Scharoun and Bryden (in press) observed 3- to 4-year-olds display the least amount of sensitivity to end-state comfort in the first session. However, they were similar to older children and adults by the second session. It can thus be suggested that, despite experience with the cup, which undoubtedly influences young children's ability to facilitate end-state comfort (Knudson et al., 2012), children are still actively learning how to solve the problem of using a cup (Thelen et al., 1993). In sum, children's early years lay the foundation for the development of motor planning skills. Therefore, exploring the environment, children become more familiar with objects, which undoubtedly influences the ability to differentiate between comfortable and uncomfortable postures, and subsequently plan according to end-state comfort.

Beginning-State Comfort in Neurotypical Development

Observations of end-state comfort are useful, as they can be used to investigate how movements are planned prior to initiation (Rosenbaum et al., 1992). In addition, researchers are beginning to investigate how end-state can be applied to a joint action paradigm, in order to better understand how one person anticipates the motor intentions of another (e.g. Gonzalez et al, 2011; Ray and Welsh, 2011). Gonzalez and colleagues (2011) investigated whether an individual will incur the cost of a movement to maximize the benefit—or beginning-state comfort—for another person. This study incorporated a toy hammer, a calculator, and a stick painted half black and half white. Participants were asked to use or place the tool individually, and, directly, or indirectly pass the tool to a confederate, who would subsequently use the tool or leave it in place. Results revealed participants facilitated beginning-state comfort for the confederate; however, did not sacrifice their own end-state comfort to do so (Gonzalez et al., 2011).

Ray and Walsh (2011) also used a joint action task to investigate whether adults select their movements to facilitate a comfortable grasp for their coactor. More specifically, participants were asked to pass a jug of water (with a handle) to a confederate. Instructions were limited, as the authors were intrigued to see the manner in which participants would pass the jug. Similar to Gonzalez and colleagues (2011), participants passed the object (jug) in a manner which facilitated a comfortable beginning-state for the recipient.

A recent study in our lab (Scanlan, Scharoun, and Bryden, 2012) investigated how object location, orientation and task interact in advanced motor planning. Using a preferential reaching paradigm, coffee mugs were arranged at three locations in peripersonal space, with handles oriented in four directions (to the left, right, away and towards the participant). Participants completed two tasks: (a) pass the mug to the researcher so they can pour a glass of water and (b) pour a glass of water and pass it to the researcher. Beginning-state comfort was deemed present if the mug was passed in a way that allowed the handle to be grasped comfortably, without further manipulation. Results revealed that regardless of the task, when the handle was positioned away from the participant, the mug was more likely to be passed in a manner which facilitated a comfortable beginning-state for the researcher. As such, the participant was willing to incur the personal cost of facilitative beginning-state comfort, because it was low. In contrast, when the handle was oriented towards the participant, the cost of facilitating beginning-state comfort was higher, as further manipulation of the cup was required; therefore, participants were less inclined to facilitate beginning-state comfort. This highlights the importance of cost minimization in motor planning (Fischer, Rosenbaum, and Jonathan, 1997).

Tasks constraints also proved influential in the participant's willingness to facilitate beginning-state comfort for the recipient. More specifically, when the handle was oriented to the left and away from the participant, more beginning-state comfort was facilitated for the recipient in the task which required the researcher to pour water. This offers some support to the suggestion that grasp postures are associated with the perceived action that the recipient will perform (Gonzalez et al., 2011). That said, this study included a limited number of participants (*N*=24); therefore, we are continuing to collect data to see whether increasing the power of our sample will influence results (Scanlan, Scharoun and Bryden, 2012).

Overall, these studies highlight that adults consider another's needs, in addition to their own, when planning joint action movements. However, to our knowledge, no one has investigated the development of beginning-state comfort to date. As such, a recent study in our lab (Scharoun and Bryden, in press) aimed to delineate how children develop motor planning skills in joint action. More specifically, at what age do children consider the beginning-state comfort of another?

Using a modified overturned glass task, neurotypical 3- to 12-year-old children and adults were asked to pick up a cup and pass it to the researcher to pour a glass of water, where cup orientation altered between right-side up and overturned. In this context, neurotypical children displayed adult-like patterns of beginning-state comfort at the age of 7, where manipulating the overturned cup highlighted developmental differences. More specifically, 3- to 4-year-olds facilitated beginning-state comfort the researcher 59.7% of the time, 5- to 6-year-olds 71.4% of the time, whereas 7- to 8-year-olds, who performed similar to older children and adults, facilitated beginning-state comfort 94.8% of the time (Scharoun and Bryden, in press).

It is well known that children are better able to relate objects to their own body, as opposed to other objects (e.g., Lockman and Ashmean, 1983; Rochat, 1988). Additionally, studies investigating motor planning skills have observed that children are more successful with self-directed movements, in comparison to other-directed movements, because the consequences of the movement are more obvious when the action is self-directed (e.g., Claxton et al., 2009; McCarty et al., 2001). This is due, in part to egocentrism, such that children do not possess the cognitive skills necessary to understand the perspectives of another (Payne and Issacs, 2012; Piaget, 1953). Observationally, the inability for young children to perceive the researchers goal was quite clear. For example, one child in the 3- to 4-year-old group manoeuvred the cup as if it were a rocket ship. It should be noted that this study was limited, as only the grip selected was analyzed (Scharoun and Bryden, in press). Future research should consider observational differences in children's approach to the task. Furthermore, as previous studies have observed discrepancies in the development of end-state comfort when comparing a 'natural' overturned glass task to a 'neutral' bar-transport task (Knudson et al., 2012), future research should investigate whether familiarity of the object influences joint action.

Overall, literature to date has revealed that sensitivity to end-state comfort and the tendency to facilitate beginning-state comfort both emerge over the course of sensory-motor development. More specifically, researchers have observed that neurotypical 7-year-olds consider another's beginning-state comfort; however, do not facilitate their own end-state comfort until the age of 9. Although the intricate details of these constructs are far from being fully delineated, observations of neurotypical development have built a strong foundation to investigate anomalies of motor development, which characterize specific developmental disabilities.

Autism Spectrum Disorder

Autism Spectrum Disorder (ASD) is one of the most common forms of neurological disorders or severe developmental disabilities of childhood (Fombonne, 2009), where it was recently reported that one in every 200 children is diagnosed (Kandel, Schwartz, Jessell, Siegelbaum, and Hudspeth, 2013). Furthermore, current estimates suggest that prevalence rates are as high as 110 per 10,000 individuals, and are rising rapidly (Kogan et al., 2009; Matson and Kozlowski, 2011). However, considerable debate exists surrounding the source of these increasing rates: changes in diagnostic criteria, new assessment tools, inaccurate diagnosis, differential research methods, increase in awareness, and cultural differences have all been implicated (see Matson and Kozlowski for a review).

Autism was first coined in 1911 by Eugen Bleuler from the Greek word *autos*, meaning *self*. The term originally referred to a basic disturbance in schizophrenia, in short, "detaching oneself from outer reality along with a relative or absolute predominance of inner life" (Bleuler, 1911, p. 304; c.f. Stotz-Ingenlath, 2000, p. 157). It was not until 1943 that Leo Kanner identified infantile autism as a unique syndrome during a case study of eleven children who shared numerous common characteristics, possessing, from birth, what he called an *extreme autistic aloneness* and an *insistence on sameness* (Folstein and Rosen-Sheidley, 2001; Kanner, 1943).

Formal diagnostic criteria for ASD emerged in 1980 with the first publication of the Diagnostic and Statistic Manual of Mental Disorders, Third Edition (DSM-III; American Psychiatric Association [APA], 1980). The most recent edition (DSM-IV-TR; APA, 2000) describes Autism as a pervasive developmental disorder by origin. Autism is a general term used to describe five unique, yet similar diagnoses encompassing ASD: Autistic Disorder, Asperger's Syndrome, Pervasive Developmental Disorder Not Otherwise Specified (Including Atypical Autism), Childhood Disintegrative Disorder and Rett's Syndrome. The newest version of the DSM is set to be released in May 2013 (APA, 2013). Proposed changes include, but are not limited to, "a complete removal of Rett's Disorder and collapsing the remaining four diagnoses into one – ASD" (Matson and Kozlowski, 2011).

Taking into consideration the spectrum which defines ASD, individuals living with this disorder range on a continuum rooted in atypical language and social development, in conjunction with repetitive patterns of behaviour before the age of three (APA, 2000). The literature (unlike DSM criteria) stereotypically denotes high- and low-functioning autism as subtypes of this disorder (e.g. Rinehart, Bradshaw, Brereton and Tonge, 2002). As such, numerous studies to date have aimed to control for IQ within ASD research. However, past research has displayed that low IQ (<80) children with ASD are typically unable and/or unwilling to complete tasks (e.g. Mandelbaum et al., 2006), thus research is typically limited to high functioning children with ASD.

While it is well documented that the core symptoms of ASD include impairments in social interaction, communicative deficits and stereotyped behaviours (APA, 2000), there is a growing body of evidence to support the presence of motor impairments in this population (e.g. Fournier, Hass, Naik, Lodha, and Cauraugh, 2010). A recent synthesis and meta-analysis argues that "motor deficits are a potential core feature of ASD" (Fournier et al., 2010, p. 1237). That said, it is suggested that, in comparison to the general population, motor impairments are more prevalent in ASD (Matson, Matson, and Beighley, 2011). Within the literature pertaining to ASD, motor impairments are typically categorized according to gross and fine motor impairment. However, special attention has been bestowed upon motor planning deficits. The remaining sections will thus introduce gross and fine motor impairments exhibited in this population. The chapter will conclude with a discussion of the current literature, which suggests impairments in end- and beginning-state comfort exemplify motor planning deficits observed in ASD.

Fine and Gross Motor Impairments in ASD

In children with ASD, both gross and fine motor impairments have been documented in the recent literature (e.g. Fournier et al., 2010; Gowen and Hamilton, 2013). Gross motor impairments include general clumsiness or ill coordinated movements (Burgoine and Wing, 1983), choreiform movements characterized by both small and large jerky motion of the arms and fingers (Jones and Prior, 1985), postural instability, hand flapping and poor performance on standardized motor function tests (Ghaziuddin and Butler, 1998; Jones and Prior, 1985; Kohen-Raz, Volkmar, and Cohen, 1992; Molloy, Dietrich, and Bhattacharya, 2003; Rapin, 1997; Rogers, Bennetto, McEvoy, and Pennington, 1996; Vilensky, Damasio, and Maurer, 1981). In addition, delayed acquisition of gross motor milestones has been reported in this population. For example, Ming, Brimacombe, and Wagner (2007) reported that children with autism were delayed in reaching major motor milestones such as walking independently, walking up steps or ramps and jumping when compared to their control counterparts (Ming, Brimacombe, and Wagner, 2007).

Hypotonia has also been documented in populations with autism. Defined as reduced resistance of the limbs during passive movements, this has been observed in both distal and proximal limbs, where researchers have noted increased joint mobility/ flexibility in the fingers and the shoulders (Ming, Brimacombe, and Wagner, 2007). Another common gross motor impairment reported in ASD is motor apraxia, defined by researchers as the inability to perform a skilled movement while possessing the desire or ability to perform them. Some examples include excessive drooling, open mouth while the individual is at rest, the inability to pucker lips and the inability to grasp tools for proper use such as using a pen, folding paper or putting together a puzzle (Ming, Brimacombe, and Wagner, 2007).

Similarly, fine motor impairments have been observed in ASD, such as impaired finger-thumb opposition during grasping. Mari, Castiello, Marks and Marrafffa (2003) observed an impaired ability to accurately sequence reach-to-grasp movements in young, low functioning children with ASD. More specifically, it was reported that when asked to employ a pincer grip and reach for a small palm-sized cube, young children with ASD displayed delayed aquisition of maximum grip aperture, increased time required to decelerate the hand prior to grasping the object as well as increased movement duration when compared to high functioning and healthy controls. These results highlight differences in movement planning and execution that exist in children with ASD (Mari et al., 2003).

Finally, it has been shown that individuals with autism exhibit challenges when controlling hand movements during the initial phases of a high accuracy pointing task.

Glazebrook and colleagues (2006) observed that adults with autism exhibited more spatial and temporal variability in the initial stages of a movement compared to their adult counterparts when asked to land on a target within a specified time frame. It is important to note that the adults with autism did not sacrifice pointing accuracy when landing on the target while exhibiting this initial variability, suggesting that this population may be unable to generate enough muscular force in the initial stages of movement tasks prior to reaching peak velocity (Glazebrook, Elliott, and Lyons, 2006).

Motor Planning in ASD

Taking the aforementioned into consideration, motor planning impairments in children with ASD have been well documented in recent years, as a number of researchers have reported significant movement impairments in the early stages of motor tasks. For example, Rinehart, Bradshaw, Brereton, and Tonge (2001) measured the ability of children with high-functioning autism (HFA), Asperger's Syndrome (AS) and a neurotypical control group to complete a reactive button depressing task—upon depression of one button, another was illuminated—as quickly as possible. It was revealed that both HFA and AS groups exhibited normal movement execution times, but delayed movement preparation times, suggesting an impairment in the ability of this population to accurately plan their movements (Rinehart et al., 2001).

In a similar study, Rinehart and colleagues (2006) utilized a choice reaction task to demonstrate that children with HFA exhibit slower movement preparation and initiation times as compared to a group of individuals with AS and their neurotypical counterparts (Rinehart et al., 2006). More recently, Dowd, McGinley, Taffe and Rinehart (2012) revealed that, when asked to perform a high accuracy pointing paradigm, children with ASD employed significantly longer movement preparation times than their neurotypical counterparts (Dowd et al., 2012). Increasing the complexity of the accuracy task, Mari and colleagues (2003) compared children with ASD and neurotypical children's ability to perform a reach-to-grasp movement while measuring the movement kinematics of the upper arm. It was revealed that children with ASD showed significant differences in movement preparation and movement execution when compared to the neurotypical control group (Mari et al., 2003).

Martineau, Schmitz, Assainte, Blank, and Barthelemy (2004) reported complementary findings, through measurement of brain activity with electroencephalography (EEG). The authors employed an anticipatory postural adjustment task of the forearms to measure Event-Related Desynchronization (ERD), an occurrence which precedes movement onset in neurotypical adult populations. Children with ASD displayed an absence of ERD both before lifting the object and upon placement of the object on their forearms. This was in contrast to ERD occurring over the right and left motor areas in neurotypical children in both the anticipation of loading and physical loading of the object onto the forearms. The authors interpreted these findings as a failure to anticipate postural and motor control in young children with ASD (Martineau et al., 2004). Overall, the aforementioned studies provide evidence to support the notion of impairment in motor planning skills in children with ASD, characterized by delayed movement preparation and execution (Dowd et al., 2012; Mari et al., 2003; Martineau et al., 2004; Rinehart et al., 2006, 2001). Analogous results have been observed in adult populations with ASD (see Glazebrook, Elliott, and Lyons, 2006; Glazebrook, Elliott, and Szatmari, 2008).

In contrast to quantifying kinematic parameters in a single step motor command, as was done in the aforementioned studies, researchers have investigated movement kinematics in multi-step movements as well (e.g., Cattaneo et al., 2007; Fabbri-Destro, Cattaneo, Boria, and Rizzolatti, 2009). Fabbri-Destro and colleagues (2009) used a three-step movement task requiring participants to lift their hand, pick up an object and place it into a container (either small or large). The two container setup forced participants to modulate the previous steps prior to finalizing the movement. It was reported that while neurotypical children would modulate the previous two steps to allow for accurate placement of the object into the container, whereas children with ASD did not modulate these steps. It was thus suggested that children with ASD approach a movement without the ability chain actions together, focusing on each step individually, rather than the movement as a whole (Fabbri-Destro et al., 2009). Additional evidence for this inability to chain actions emerged upon measuring electromyography (EMG) of the mylohyoid (MH) muscle of the mouth. When asked to grasp a food item and bring it towards the mouth, neurotypical children activated the MH much earlier in the movement sequence than did children with ASD. The MH activated upon grasping a food item in TD children, whereas the MH muscle only became active upon bringing the food towards the mouth in children with ASD. It can thus be suggested that children with ASD display an impaired ability to chain actions together into one cohesive movement (Cattaneo et al., 2007).

End- and Beginning-State Comfort in ASD

It is well understood that ASD is characterized by impairments in the ability to plan, organize, and coordinate movements (e.g., Glazebrook et al., 2006, 2008; Gowen and Hamilton, 2013). That said, motor planning abilities in individuals with ASD have also been assessed using to the end-state comfort effect. However, literature to date reports both impairment and similar performance when comparing children with ASD to their neurotypical counterparts, thus highlighting controversial results. Recent investigations have also aimed to extend end-state comfort to joint action paradigms, questioning whether children (Scharoun and Bryden, in press) and adults (Gonzalez, Glazebrook, Studenka and Lyons, 2013) consider the beginning-state comfort of another when passing objects in motor planning tasks.

Hughes (1996) was the first to investigate end-state comfort in 36 children with autism (according to DMS-III-R; APA, 1987). As a means of comparison, 24 children attending schools for children with Moderate Learning Disabilities (MLD) were included, where Hughes (1996) noted that MLD “is a British education term that is approximately equivalent to the North American term ‘mental retardation’ or ‘mental handicap’” (p. 101). Finally, 28 neurotypical children were included as controls. Participants were separated into two subgroups according to nonverbal mental age, according to the Matrices task of the British Abilities Scale (Elliott, Murray, and Pearson, 1983). Participants were asked to place the end of a two-coloured bar into one of two disks, where the end posture of the hand upon insertion was examined. End-state comfort was defined as a thumb-up grip on the bar. It was observed that children with autism were less sensitive to end-state comfort than both the group of children with MLD and neurotypical pre-school children (Hughes, 1996). Hughes (1996) thus suggested planning problems at the level of motor control are evident in ASD.

However, more recent investigations have observed no group differences when comparing children with ASD to their neurotypical counterparts. Hamiltom, Brindley, and Frith (2007) investigated 23 children with ASD (independent clinical diagnosis) and 31 neurotypical counterparts. Verbal mental age was established using the British Picture Vocabulary Scale (Dunn, Dunn, Whetton, and Burley, 1997). Using a horizontal bar transport task, the authors observed comparable motor planning skills in both groups of children. Van Swieten and colleagues (2010) explored the ability of children with autism (diagnosed using the Autism Diagnostic Interview – Revised; Lord, Rutter, and Le Couteur, 1994), developmental coordination disorder (DCD; by DSM-IV criteria; 6- to 8-year-olds and 9-to 13-year-olds) and their neurotypical counterparts (5- to 8-, 9- to 14- and 19- to 32-year-olds) to rotate a cylinder (clockwise or anticlockwise) while maintaining an end-state comfortable position. Children with DCD favoured a comfortable initial grip posture while lacking a comfortable final grip posture. Conversely, neurotypical children and children with autism achieved the same percentage of end-state comfortable postures. This was hypothesized to be due to low demands of the task, which may have elicited deficits in an ASD population, due to impairments in executive functions (van Swieten et al., 2010).

A recent study in our lab (Scharoun and Bryden, 2013) aimed to further elucidate the differences in motor planning skills between 5- to 12-year-old children with ASD (independent clinical diagnosis) and their neurotypical counterparts. Using the same methods as Scharoun and Bryden (in press), participants were asked to pick up a cup and (a) pour a glass of water or (b) pass it to the researcher to pour a glass of water; therefore enabling observations of both sensitivity to end-state comfort and facilitation of beginning-state comfort for the recipient. Cup placement was altered between right-side up and overturned. Within this context, children with ASD displayed significantly less evidence of end-state comfort and beginning-state comfort than their neurotypical counterparts. This was exaggerated when manipulating the inverted cup, where children were challenged to re-orient the cup. Hence, the inverted cup presented a more challenging movement, where it is suggested that complexity demands associated with an inverted cup may hinder end-state comfort in children, considering more degrees of freedom must be controlled to solve the action (Weigelt and Schack, 2010). Furthermore, when pouring water, children are required to skillfully coordinate the two hands; therefore splitting a child's attention (Weigelt and Schack, 2010). As such, motor impairments in children with ASD were highlighted with constraining task elements.

Research has noted that children with ASD possess deficits in sequencing abilities (Bernstein, 1967; Gobet et al., 2001; Graybiel, 1998; Hughes, 1996) and are restricted in their understanding of the consequences associated with one's own actions (Hughes, 1996; Poulton, 1957; von Hofsten, 1990). Children with ASD did not display differences in grip selection when manipulating the cup oriented right-side up, in comparison to the overturned cup (Scharoun and Bryden, 2013). This offers support to the suggestion that children with ASD may not consider the final goal when planning their movements (Fabbri-Destro et al., 2009; Gowen and Hamilton, 2013; Hughes, 1996). Once an action plan is constructed, the cost of deviating from the original plan is both difficult and costly for children with ASD (Hill, 2004). Consequently, children with ASD displayed awkward grip selections when manipulating the overturned glass, and were unable to facilitate beginning-state comfortable postures for a recipient, within this motor planning task (Scharoun and Bryden, 2013).

Interestingly, one study has revealed that adults with ASD do not display the same deficits as children with ASD, when observing end- and beginning-state comfort. Using three common tools (a wooden toy hammer, a stick, and a calculator), Gonzalez and colleagues (2013) asked participants to pass a tool to a confederate, who would subsequently place the tool down, or use the tool. Both neurotypical adults and adults with ASD facilitated beginning-state comfort, thus passing the tool to accommodate for the confederate's action (Gonzalez et al., 2013). However, adults with ASD were more variable in performance. The authors thus suggested that motor planning abilities are not consistent across individuals with ASD. Therefore, measuring performance during joint action can help to elucidate patterns of motor planning abilities of individuals with ASD when involved in interpersonal relations (Gonzalez et al., 2013). Future research is thus required, to further delineate trends in development from early childhood to adulthood.

Currently, the degree to which individuals with ASD can plan a movement according to end- and beginning-state comfort remains inconclusive. Researchers have been left questioning the underlying mechanisms, therefore solidifying the need further investigation. Some researchers have investigated the role of impaired action understanding (Hamilton, Brindley, and Frith, 2007), where perhaps the participant cannot understand the intent of the task and therefore cannot complete the task successfully. It has also been hypothesized that there are problems upstream of the motor system such as the ventral and dorsal visual streams of the neural system, where individuals with ASD have exhibited impairments in accurately perceiving environmental stimuli (Bölte, Holtmann, Poustka, Scheurich and Schmidt, 2007; Happe, 1996; Mitchell, Mottron, Soulieres and Ropar, 2010). Finally, it has been proposed that issues lie within the Mirror Neuron System (MNS), which becomes active in healthy adults when observing an action and while also performing an action (Rizzolatti and Craighero, 2004). See Fournier and colleagues (2010) and/or Gowen and Hamilton (2013) for a review.

Summary and Conclusion

Rosenbaum and colleagues (1990) coined "end-state comfort" to describe participants' approach to a bar transport task. Adults were observed selecting an uncomfortable initial posture (underhand supination grip; thumb up) in order to end their movement with a neutral, comfortable posture (overhand pronation grip; thumb-down). End-state comfort has since been used to investigate behaviours associated with advanced motor planning (e.g. Cohen and Rosenbaum, 2004; Haggard, 1998; Rosenbaum, van Heugten and Caldwell, 1996; Rosenbaum and Jorgensen, 1992; Rosenbaum et al., 1992; Short and Cauraugh, 1996; Weigelt, Kunde, and Prinz, 2006). The bulk of research to date has relied on adult participants; however, a handful of studies have investigated end-state comfort in neurotypical development (e.g. Adalbjornsson, et al., 2008; Manoel and Moreira, 2005; Scharoun and Bryden, in press; Smyth and Mason, 1997; Stö ckel et al., 2011; Thibaut and Toussaint, 2010; Weigelt and Schack, 2010).

The developmental literature highlights the link between movement capabilities and cognitive development; therefore suggesting end-state comfort is a motor phenomenon acquired with age. Young children lack sensitivity to end-state comfort, thus demonstrating a variety of grip selection strategies. Children show significant improvements in sensitivity within testing sessions as a function of age, as they use motor skills to explore the environment and learn to overcome the degrees of freedom associated with specific movements. Children are more successful with self-directed movements; however, sensitivity to end-state comfort in other-directed movements emerges as children garner the ability to visualize the perspectives of others, and consider the consequences of their actions. Overall, as mentioned previously, "motor planning works as a blind watchmaker, with actions reflecting a previous history of motor evolution where useful actions have survived and less useful ones have perished" (van Swieten et al., 2010, pp. 498). Children's early years lay the foundation for the development of motor planning skills. Exploring the environment, through trial-and-error, children learn the affordances of objects, become more familiar with comfortable and uncomfortable postures, and learn to plan according to end-state comfort. Adult-like patterns of end-state comfort have thus been observed in 9-year-old children.

Recent investigations have expanded upon Rosenbaum and colleagues (1990) concept of end-state comfort with application to joint action paradigms (e.g. Gonzalez et al., 2011; Ray and Welsh, 2011; Scharoun and Bryden, in press). Although limited, research to date has observed that participants as young as 7-years-old ensure an object is passed in a manner that facilitates beginning-state comfort for the recipient (Gonzalez et al., 2011; Ray and Welsh, 2011; Scharoun and Bryden, 2013).

Although end- and beginning-state comfort in neurotypical development have yet to be fully delineated, observations of these constructs have built a strong foundation to investigate anomalies of motor development, which characterize developmental disabilities, such as ASD. The ASD literature has demonstrated inconclusive results, such that some (e.g., Hughes, 1996; Scharoun and Bryden, 2013) have observed children with ASD demonstrate significantly less end- and beginning-state comfort than their neurotypical counterparts. However, others (e.g., Hamilton et al., 2007; van Swieten et al., 2010) have noted similar performance in both groups of children.

Overall, recent investigations have revealed that, in neurotypical development, adult-like patterns of beginning-state comfort emerge at the age of 7; however, children do not facilitate their own end-state comfort until the age of 9. Using neurotypical development as a foundation to investigate developmental disability, research to date has displayed inconclusive findings. Future research is required in both neurotypical populations and developmental disabilities, such as ASD, to elucidate trends in end and beginning-state comfort from early childhood to adulthood.

Implications and Future Directions

It is well known that "neurological diseases exact an exorbitant health cost on our population" (Ajemian and Hogan, 2010, p. 337). In addition to financial burdens, families and caregivers of children with ASD face challenges in various aspects of daily life, including social (Gray, 1993), health (Allik, Larsson and Smedge, 2006) and overall life/lifestyle (Sen and Yurtsever, 2007). To mitigate these costs, a better understanding of the neurological foundations of these diseases is required (Ajeman and Hogan, 2010). It was recently suggested that motor coordination deficits are a cardinal feature of ASD (Fournier et al., 2010). Furthermore, that future work should focus on motor planning, as these skills play a significant role in the underlying causes of impairment in ASD (Gowen and Hamilton, 2013). Not only does this require further research with special populations, such as ASD, but also in neurotypical development, to help solidify a foundation for comparison.

References

American Psychiatric Association. (1980). Diagnostic and statistical manual of mental disorders (3rd ed.). Washington, DC: Author.

American Psychiatric Association. (2013). DSM-5 Retrieved February 20, 2013, from www.dsm-5.org.

American Psychiatric Association. (2000). Diagnostic and statistical manual of mental disorders (4th ed. – text revision). Washington, DC: Author.

Adalbjornsson, C. F., Fischman, M. G., and Rudisill, M. E. (2008). The end-state comfort effect in young children. *Research Quarterly for Exercise and Sport,* 79(1), 36-41.

Ajemian, R., and Hogan, N. (2010). Experimenting with theoretical motor neuroscience. *Journal of Motor Behavior*, 42(6), 333-342.

Allik, H., Larsson, J., and Smedje, H. (2006). Health-related quality of life in parents of school-age children with Asperger syndrome or high functioning autism. *Health and Quality of Life Outcomes*, 4, 1-8.

Bernstein, N. A. (1967). The coordination and regulation of movements. Oxford: Pergamon Press.

Bleuler, E. (1911), 'Dementia praecox oder Gruppe der Schizophrenien', In: G. Aschaffenburg (ed.), *Handbuch der Psychiatrie. Spezieller Teil. 4. Abteilung, 1.Hälfte.* Leipzig und Wien: Franz Deuticke.

Bö lte, S., Holtmann, M., Poustka, F., Scheurich, A., and Schmidt, L. (2007). Gestalt perception and local-global processing in High-Functioning Autism. *Journal of Autism and Developmental Disorders,* 37, 1493–1504.

Bryden, P. J., Roy, E. A., and Spence, J. (2007). An observational method of assessing handedness in children and adults. *Developmental Neuropsychology*, *32*(3), 825-846.

Burgoine, E., and Wing, L. (1983). Identical triplets with Asperger's syndrome. *British Journal of Psychiatry,* 143, 261-265.

Carruth, B.R., and Skinner, J.D. (2002). Feeding behaviors and other motor development in healthy children (2-24 months). *Journal of the American College of Nutrition*, 21(2), 88-96.

Cattaneo, L., Fabbri-Destro, M., Boria, S., Pieraccini, C., Monti, A., Cossu, G., and Rizzolatti, G. (2007). Impairment of actions chains in autism and its possible role in intention understanding. *Proceedings of the National Academy of Sciences*, 104(45),17825–17830.

Claxton, L. J., McCarty, M. E., and Keen, R. (2009). Self-directed action affects planning in tool-use tasks with toddlers. *Infant Behaviour Development*, 32(2), 230-233.

Cohen, R. G., and Rosenbaum, D. A. (2004). Where grasps are made reveals how grasps are planned: Generation and recall of motor plans. *Experimental Brain Research*, 157, 486–495.

Cruse, H. (1986). Constraints for joint angle control of the human arm. *Biological Cybernetics*, 54, 125–132.

Cruse, H., Wischmeyer, E., Brüwer, M., Brockfeld, P., and Dress, A. (1990). On the cost functions for the control of the human arm movement. *Biological Cybernetics*, 62, 519–528.

Dowd, A. M., McGinley, J. L., Taffe, J. R., and Rinehart, N. J. (2012). Do planning and visual integration difficulties underpin motor dysfunction in Autism? A Kinematic Study of Young Children with Autism. *Journal of Autism and Developmental Disorders*, 42(8), 1539-1548.

Dunn, L. M., Dunn, L. M., Whetton, K., and Burley, J. (1997). *British picture vocabulary scale* (2nd ed.). Windsor: NFER-NELSON.

Elliot, C., Murray, D., and Pearson, L. (1983). *British Abilities Scales.* Windsor: NFER Nelson.

Fabbri-Destro, M., Cattaneo, L., Boria, S., and Rizzolatti, G. (2009). Planning actions in autism. *Experimental Brain Research*, 192, 521–525.

Fischer, M., Rosenbaum, D., and Jonathan, V. (1997). Speed and sequential effects in reaching. *Journal of Experimental Psychology: Human Perception and Performance, 23*(2), 404-428.

Fischer, M. H., Rosenbaum, D. A., and Vaughan, J. (1997). Speed and sequential effects in reaching. *Journal of Experimental Psychology: Human Perception and Performance*, 23, 404–428.

Fischman, M. G. (1997). End-state comfort in object manipulation [Abstract]. *Research Quarterly for Exercise and Sport*, 68 (Suppl.), A-60.

Folstein, S. E., and Rosen-Sheidley, B. (2001). Genetics of autism: complex aetiology for a heterogeneous disorder. *Nature Reviews Genetics*, *2*(12), 943-955.

Fombonne, E. (2009). Epidemiology of pervasive developmental disorders.*Pediatric research*, 65(6), 591-598.

Fournier, K. A., Hass, C. J., Naik, S. K., Lodha, N., and Cauraugh, J. H. (2010). Motor coordination in Autism Spectrum Disorders: A synthesis and meta-analysis. *Journal of Autism and Developmental Disorders*, 40, 1227-1240.

Ghaziuddin, M., and Butler, E. (1998). Clumsiness in autism and Apserger syndrome: A further report. *Journal of Intellectual Disability Research, 42*43-48.

Glazebrook, C. M., Elliott, D., and Lyons, J. (2006). A kinematic analysis of how young adults with and without Autism plan and control goal-directed movements. *Motor Control*, 10(3), 244–264.

Glazebrook, C. M., Elliott, D., and Szatmari, P. (2008). How do individuals with autism plan their movements? *Journal of Autism and Developmental Disorders*, 38(1), 114–126.

Gobet, F., Lane, P. C. R., Croker, S., Cheng, P. C-H., Jones, G., Oliver, I., and Pine, J.M. (2001). Chunking mechanisms in human learning. *TRENDS in Cognitive Sciences*, 5(6), 236-243.

Gonzalez, D.A., Glazebrook, C.M., Studenka, B.E., and Lyons, J.L. (2013). *Motor intentions with another person: Do individuals with Autism Spectrum Disorders plan ahead?* Manuscript submitted for publication.

Gonzalez, D. A., Studenka, B. E., Glazebrook, C. M., and Lyons, J. L. (2011). Extending end-state comfort effect: Do we consider the beginning state comfort of another? *Acta Psychologica*, 136, 347-353.

Gowen, E., and Hamilton, A. (2013). Motor abilities in autism: A review using a computational context. *Journal of Autism and Developmental Disorders*, 42(2), 323-344.

Gray, D.E. (1993). Perceptions of stigma: The parents of autistic children. *Sociology of Health and Illness*, 15(1), 102–120.

Graybiel, A. M. (1998). The basal ganglia and chunking of action repertoires. *Neurobiology of Learning and Memory*, 70,119–136.

Haggard, P. (1998). Planning of action sequences. *Acta Psychologica*, 99, 201–215.

Hamilton, A., Brindley, R.M., and Frith, U. (2007). Imitation and action understanding in autistic spectrum disorders: How valid is the hypothesis of a deficit in the mirror neuron system? *Neuropsycohlogia*, 45, 1859-1868.

Happe, F. G. E. (1996). Studying weak central coherence at low levels: Children with Autism do not succumb to visual illusions. A research note. *Journal of Child Psychology and Psychiatry*, 37(7), 873-877.

Hill, E. L. (2004). Executive dysfunction in autism. *Trends in cognitive sciences*, *8*(1), 26-32.

Herbort, O., and Butz, M. V., (2011). Habitual and goal-directed factors in (everyday) object handling. *Experimental Brain Research*, 213, 371-382.

Hughes, C. (1996). Brief report: Planning problems in autism at the level of motor control. *Journal of Autism and Developmental Disorders*, 26(1), 99-107.

Jones, V., and Prior, M. (1985). Motor imitation abilities and neurological signs in autistic children. *Journal of Autism and Developmental Disorders*, 15(1), 37–46.

Jordan, M. I., and Rosenbaum, D. A. (1989). Action. *Foundations of cognitive science*, 727-767.

Kandel, E.R., Schwartz, J.H., Jessell, T.M., Siegelbaum, S.A., and Hudspeth, A.J. (2013). *Principles of Neural Science* (5th Ed.). New York, NY: McGraw Hill Medical.

Kanner, L. (1943). Autistic disturbances of affective contact. *Nervous child*,*2*(3), 217-250.

Kent, S. W., Wilson, A. D., Plumb, M. S., Williams, J. H., and Mon-Williams, M. (2009).
Immediate movement history influences reach-to-grasp action selection in children and adults. *Journal of Motor Behavior*, *41*(1), 10-15.

Knudson, B., Henning, A., Wunsch, K., Weigelt, M., and Aschersleben, G. (2012). The end-state comfort effect in 3- to 8-year-old children in two object manipulation tasks. *Frontiers in Psychology*, 3, 445, 1-10.

Kogan, M. D., Blumberg, S. J., Schieve, L. A., Boyle, C. A., Perrin, J. M., Ghandour, R. M., ... and van Dyck, P. C. (2009). Prevalence of parent-reported diagnosis of autism spectrum disorder among children in the US, 2007.*Pediatrics*, 124(5), 1395-1403.

Kohen-Raz, R., Volkmar, F. R., and Cohen, D. J. (1992). Postural control in children with autism. *Journal of Autism and Developmental Disorders*, 22(3), 419–432.

Lockman, J. J., and Ashmead, D. H. (1983). Asynchronies in the development of manual behavior. *Advances in Infancy Research*, 2, 113-136.

Lord, C., Rutter, M., and Le Couteur, A. (1994). Autism Diagnostic Interview-Revised: A revised version of a diagnostic interview for caregivers of individuals with possible pervasive developmental disorders, *Journal of Autism and Developmental Disorders, 24,* 659–685.

Mandelbaum, D. E., Stevens, M., Rosenberg, E., Wiznitzer, M., Steinschneider, M., Korey, S. R., ... and Rapin, I. (2006). Sensorimotor performance in school-age children with autism, developmental language disorder, or low IQ. *Developmental Medicine and Child Neurology*, *48*(1), 33-39.

Manoel, E. J., and Moreira, C. R. P. (2005). Planning manipulative hand movements: Do young children show the end-state comfort effect? *Journal of Human Movement Studies*, 49, 93-114.

Mari, M., Castiello, U., Marks, D., Marraffa, C., and Prior, M. (2003). The reach-to-grasp movement in children with autism spectrum disorder. *Philosophical Transactions of the Royal Society B: Biological Sciences*, 358(1430), 393–403.

Martineau, J., Schmitz, C., Assaiante, C., Blanc, R., and Barthelemy, C. (2004). Impairment of a cortical event-related desynchronisation during a bimanual load-lifting task in children with autistic disorder. *Neuroscience Letters*, 367(3), 298–303.

Matson, J. L., and Kozlowski, A. M. (2011). The increasing prevalence of autism spectrum disorders. *Research in Autism Spectrum Disorders*, *5*(1), 418-425.

Matson, M.L., Matson, J.L., and Beighley, J.S. (2011). Comorbidity of physical and motor problems in children with autism. *Research in Developmental Disabilities*, 32, 2304-2308.

McCarty, M. E., Clifton, R. K., and Collard, R. R. (1999). Problem solving in infancy: The emergence of an action plan. *Developmental Psychology*, 35, 1091-1101.

McCarty, M. E., Clifton, R. K., and Collard, R. R. (2001). The beginnings of tool use by infants and toddlers. *Infancy*, 2(2), 233-256.

Ming, X., Brimacombe, M., and Wagner, G. C. (2007). Prevalence of motor impairment in autism spectrum disorders. *Brain and Development*, 29(9), 565–570.

Mitchell, P., Mottron, L., Souliéres, I.,and Ropar, D. (2010). Susceptibility to the Shepard Illusion in Participants with Autism: Reduced Top-Down Influences Within Perception? *Autism Research 3*: 113–119.

Molloy, C. A., Dietrich, K. N., and Bhattacharya, A. (2003). Postural stability in children with autism spectrum disorder. *Journal of Autism and Developmental Disorders*, 33(6), 643–652.

Payne, G. and Issacs, L. (2012). Human motor development: A lifespan approach (8th ed.). New York, NY: McGraw Hill.

Piaget, J. (1953). *The origin of intelligence in the child.* London: Routledge and Kegan Paul Ltd.

Poulton, E. C. (1957). On prediction in skilled movements. *Psychological Bulletin, 54,* 467-478.

Rapin, I. (1997). Autism. *New England Journal of Medicine*, 337(2), 97–104.

Ray, M. and Welsh, T. M. (2011). Response selection during a joint action task. *Journal of Motor Behavior*, 43(4), 329-332.

Rinehart, N. J., Bellgrove, M. A., Tonge, B. J., Brereton, A. V., Howells-Rankin, D., and Bradshaw, J. L. (2006). An examination of movement kinematics in young people with high-functioning autism and Asperger's disorder: Further evidence for a motor planning deficit. *Journal of Autism and Developmental Disorders*, 36(6), 757–767.

Rinehart, N. J., Bradshaw, J. L., Brereton, A. V., and Tonge, B. J. (2002). A clinical and neurobehavioural review of high-functioning autism and Asperger's disorder. *Australian and New Zealand Journal of Psychiatry*, *36*(6), 762-770.

Rinehart, N. J, Bradshaw, J. L., Moss, S. A, Brereton, A. V., and Tonge, B. J. (2001). A deficit in shifting attention present in high functioning autism but not Asperger's disorder. *Autism*, 5,67–80.

Rizzolatti, G., and Craighero, L. (2004). The mirror-neuron system. *Annual Review of Neuroscience*, 27, 169-192.

Rochat, P. Self-perception and action in infancy. (1998). *Experimental Brain Research,* 123, 102–109.

Rogers, S. J., Bennetto, L., McEvoy, R., and Pennington, B. F. (1996). Imitation and pantomime in high-functioning adolescents with autism spectrum disorders. *Child Development*, 67(5), 2060–2073.

Rosenbaum, D. A., Engelbrecht, S. E., Bushe, M. M., and Loukopoulos, L. D. (1993).

Knowledge model for selecting and reproducing reaching movements. *Journal of Motor Behaviour*, 25, 217–227.

Rosenbaum, D. A., and Jorgensen, M. J. (1992). Planning macroscopic aspects of manual control. *Human Movement Science, 11*(1), 61-69.

Rosenbaum, D. A., Marchak, F., Barnes, H. J., Vaughan, J., Slotta, J. D., and Jorgensen, M. J. (1990). Constraints for action selection: Overhand versus underhand grips. In M. Jeannerod (Ed.), Motor representation and control, Attention and Performance XIII (pp. 321–342). Hillsdale, NJ: Erlbaum.

Rosenbaum, D. A., Loukopoulos, L. D., Meulenbroek, R. G., Vaughan, J., and Engelbrecht, S. E. (1995). Planning reaches by evaluating stored postures. *Psychological Review*, 102, 26–67.

Rosenbaum, D. A., Meulenbroek, R. J., Vaughan, J., and Jansen, C. (2001). Posture-based motion planning: Applications to grasping. *Psychological Review*, 108, 709–734.

Rosenbaum, D. A., van Heugten, C. M., and Caldwell, G. E. (1996). From cognition to biomechanics and back: The end-state comfort effect and the middle-is-faster effect. *Acta Psychologica*, *94*(1), 59-85.

Rosenbaum, D. A., Vaughan, J., Barnes, H. J., and Jorgensen, M. J. (1992). Time course of movement planning: selection of handgrips for object manipulation. *Journal of Experimental Psychology: Learning, Memory, and Cognition*, *18*(5), 1058.

Scanlan, K. A., Scharoun, S. M., and Bryden, P. J. (November 2012). Preferential reaching and beginning-state comfort interactions: Implications for advanced motor planning. Poster presented at the Canadian Society for Psychomotor Learning and Sport Psychology, Halifax, NS.

Scharoun, S. M., and Bryden, P. J. (in press). The development of end- and beginning-state comfort in a cup manipulation task. *Developmental Psychobiology*. DOI: 10.1002/dev.21108.

Scharoun, S. M., and Bryden, P. J. (2013). *End- and beginning-state comfort in children with Autism Spectrum Disorders and their neurotypical counterparts*. Manuscript submitted for publication.

Sen, E., and Yurtsever, S. (2007). Difficulties experienced by families with disabled children.*Journal for Specialists in Pediatric Nursing*, 12(4), 238–252.

Short, M. W., and Cauraugh, J. H. (1996). Planning macroscopic aspects of manual control: End-state comfort and point-of-change effects. *Acta Psychologica*, 96, 133−147.

Smyth, M. M., and Mason, U. C. (1997). Planning and execution of action in children with and without developmental coordination disorder. *Journal of Child Psychiatry*, 38, 1023–1037.

Stöckel, T., Hughes, C. M., and Schack, T. (2011). Representation of grasp postures and anticipatory motor planning in children. *Psychological Research*, 76(6), 768-776.

Stotz-Ingenlath, G. (2000). Epistemological aspects of Eugen Bleuler's conception of schizophrenia in 1911. *Medicine, Health Care and Philosophy*,*3*(2), 153-159.

Thelen, E., Corbetta, D., Kamm, K., Schneider, K., and Zernicke, R. (1993). The transition to reaching: Mapping intention and intrinsic dynamics. *Child Development*, 64(4), 1058-1098.

Thibaut, J. P., and Toussaint, L. (2010). Developing motor planning over ages. *Journal of Experimental Child Psychology*, 105, 116-129.

van Bergen, E., van Swieten, L. M., Williams, J. H., and Mon-Williams, M. (2007). The effect of orientation on prehension movement time. *Experimental Brain Research*, *178*(2), 180-193.

van Swieten, L. M., van Bergen, E., Williams, J. H., Wilson, A. D., Plumb, M. S., Kent, S. W., and Mon-Williams, M. A. (2010). A test of motor (not executive) planning in developmental coordination disorder and autism. *Journal of Experimental Psychology: Human Perception and Performance*, *36*(2), 493.

Vilensky, J. A., Damasio, A. R., and Maurer, R. G. (1981). Gait disturbances in patients with autistic behavior: A preliminary study. *Archives of Neurology*, 38(10), 646–649.

von Hofsten, C. (1990). A perception-action perspective on the development of manual movements. In J. Long and A. Baddeley (Eds.), *Attention and Performance* (Vol. 13, pp. 739-762). Hillsdale, NJ: Erlbaum.

Weigelt, M., Kunde, W., and Prinz, W. (2006). End-state comfort in bimanual object manipulation. *Experimental Psychology*, 53, 143−148.

Weigelt, M., and Schack, T. (2010). The development of end-state comfort planning in preschool children. *Experimental Psychology*, 57(6), 476-482.

In: Motor Behavior and Control: New Research
Editors: Marco Leitner and Manuel Fuchs
ISBN: 978-1-62808-142-8

Chapter 2

The Role of Motor Imagery in Action Planning: Implications for Developmental Research

***Carl Gabbard*[1], *Priscila Caçola*[2] *and Jihye Lee*[1]**
[1]Texas A&M University, Texas, US
[2]University of Texas at Arlington, Texas, US

Abstract

Motor imagery is a widely used experimental paradigm for the study of cognitive aspects of action planning and control in adults. Underscoring that interest is the assertion that motor imagery provides a window into the process of action representation. These notions complement internal modeling theory suggesting that such representations allow predictions (estimates) about the mapping of the self to parameters of the external world; processes that enable successful planning and execution of action. Those observations have drawn the attraction of developmentalists that work with typically developing children and special populations. This chapter defines motor imagery and its link to mental representation, internal modeling, embodied cognition, and foremost, action planning. Included is a selection of recent work with typically developing children and special populations. The merits of this review are associated with the apparent increasing attraction for studying and using motor imagery to understand the developmental aspects of action processing in children.

Introduction

This chapter presents information on the concept of motor imagery and its role in action planning. To this end, we briefly discuss the concepts of mental representation, internal modeling, motor imagery, motor cognition, and the connection to embodied cognition. And lastly, we argue that use of motor imagery has the potential to promote greater insight to the development of action planning in children.

Motor Imagery

Motor imagery (MI) is defined as an internal rehearsal or reenactment of movements from a first-person perspective without any overt physical movement. From another perspective, MI, also known as kinesthetic imagery, is an active cognitive process during which the representation of a specific action is internally reproduced in working memory without any overt motor output (Decety & Grezes, 1999). In addition to the reasonable case that MI is a reflection of action representation and motor planning, studies have found that there is a high correlation between real and simulated movements (e.g., Sharma, Jones, Carpenter, & Baron, 2008; Young, Pratt, & Chau, 2009). Complementing those findings is the idea that motor control and motor simulation states are functionally equivalent (Jeannerod, 2001; Kunz, Creem-Regehr, & Thompson, 2009; Lorey et al., 2010). Also known as 'simulation theory', this idea postulates that mental (covert) and executed (overt) actions rely on similar motor representations. Such research has drawn the interest of (for example) clinicians working to stimulate neuromuscular pathways. For example, with stroke patients, injured athletes, and those with cerebral palsy, imagery is considered an attractive means to access the motor network and restore motor function without actual overt action. Also, a relatively large body of literature indicates that sport psychologists use mental rehearsal as a means to facilitate practice and improve performance in athletes (see reviews by Munzert, Lorey, & Zentgraf, 2009 and Sharma et al., 2009). Furthermore, evidence has been reported showing that MI follows the basic tenets of Fitts' Law (Solodkin, Hlustik, Chen, & Small, 2004; Stevens, 2005). That is, simulated movement duration like actual movement, decreases with increasing task complexity. As opposed to visual imagery, defined as the internal enactment or reenactment of perceptual experiences (Barsalou, 2008), neuroimaging and neuropsychological studies indicate that MI is more affected by biomechanical (kinesthetic) constraints that are commonly associated with action processing (see review by Pelgrims, Andres, & Olivier, 2005; Stevens, 2005).

Predicting Outcome and Consequences

One of the interesting hypothesized features of MI is its role in the prediction of one's actions (e.g., Kunz et al., 2009; Lorey et al., 2010). One of the important aspects of an action plan is the ability to predict the outcome and consequences of intended actions. Suddendorf and Moore (2011) note, "The ability to imagine future events is an essential part of human cognition" (p. 295). Imagery allows us to generate specific predictions based upon past experience and allows us to answer 'what if' questions by making explicit and accessible the likely consequences of a specific action. Mental simulations generate knowledge about specific past events, and therefore make specific predictions (Moulton & Kosslyn, 2009). Imagining an action can serve several useful goals to that endeavor. According to Bourgeois & Coello (2009), motor representation can be viewed as a component of a predictive system, which includes a neural process that simulates through MI the dynamic behavior of the body in relation to the environment. This line of reasoning presents interesting developmental issues associated with the child's cognitive understanding of environmental (perceptual) information and consequences, and one's physical capabilities. Based on a recent review by

Springer, Hamilton, and Cross (2012), the importance of motor experience for prediction is well documented in adults. We would add that much less is known about children.

This general line of reasoning supports the notion of internal (forward) modeling described in the next section.

Mental (Action) Representation

The nature of mental representation is a central issue for understanding cognitive and motor development across the lifespan. The concept has been cast from several perspectives. In general, the term is used as a construct of neuro- and cognitive science involving cognitive states and processes constituted by the occurrence, transformation and storage of information-bearing structures (representations) of one kind or another (Stanford Encyclopedia of Philosophy, 2008). From another perspective, it is an internal cognitive construct that represents external reality. As presented here, imagery is a form of and key modality for the creation of representations (see review by Kosslyn, Thompson, & Ganis, 2006).

Motor programming theory in general suggests that an integral component in an effective outcome is an adequate action representation of the movements. This view contends that action representation is a key feature of an internal forward model, which is a neural system that *simulates* the dynamic behavior of the body in relation to the environment (Schubotz, 2007; Wolpert, 1997). This theory proposes that internal models make *predictions* (estimates) about the mapping of the self to parameters of the external world; processes that enable successful planning and execution of action. These representations are hypothesized to be an integral part of action planning (Caeyenberghs, Roon, Swinnen, & Smits-Engelsman, 2009; Molina, Tijus, & Jouen, 2008). Skoura, Vinter, and Papaxanthis (2009) suggest that improvement in action representation during childhood may be due to refinement of internal models. Complementing the forward model idea and central to the present discussion is the widely acknowledged proposition that simulation in the form of *motor imagery provides a window into the process of action representation*; that is, it reflects an internal action representation (Jeannerod, 2001; Munzert et al., 2009).

Developmental Studies with Typically Developing Children

Although there is no direct evidence that infants use MI, studies do provide a rather convincing case that infants take into account situational constraints when planning and executing actions (e.g., Csibra, Bírób, Koósc, & Gergely, 2003; Sommerville, Woodward, & Needham, 2005; Willatts, 1999). These studies suggest that infants show some understanding of human actions, thus seem to have a representation of such. For example, it appears that infants understand uncompleted actions, and are able to complete perceived actions. Obviously, the inherent limitation of such studies is the problem with determining the level of cognitive processing occurring. That is, infants cannot adequately communicate verbal judgments. Therefore, the difficulty with differentiating between cognitive and neuromotor processes in the programming of movements is compounded. Arguably, use of reaching contacts, forward lean, looking time, and gaze, are not ideal for addressing the issue of motor cognition.

A review of the literature indicates that by 5 years of age, the ability to effectively create and use MI to represent movement is present (e.g., Caeyenberghs, Tsoupas, Wilson, & Smits-Engelsman, 2009; Frick, Daum, Wilson, & Wilkening, 2009; Molina et al., 2008; Smits-Engelsman & Wilson, in press). From the work just referenced, a number of innovative tactics have been created to examine the ability of children to mentally represent action via use of MI. Two of the more popular methods have been mental rotation of different hand positions and the chronometry paradigm. With the typical hand rotation task, participants are asked to judge whether a hand visually viewed in an unusual orientation is a right or left limb; reaction time is commonly recorded also. The usual strategy is to 'imagine' one's own body. Chronometry involves the comparison of simulated and actual movements. That is, the correspondence between the time-course of the participant's imagined and executed actions. This tactic follows the premise that there is a functional equivalence (relationship) between MI and execution. With such studies, an array of movement tasks have been used with children; namely: walking, grasping, moving a puppet, pointing, figure drawing, and time to complete an obstacle course.

A few specific studies that demonstrate innovation are worth noting. Molina and colleagues (2008) used a chronometry paradigm with children 5- to 7 years to compare movement duration of actually moving (walking) a puppet to a location, and imagining executing the same action. Movement durations for actual and simulated displacements were obtained in two conditions, where either no information was provided about the weight of the puppet to be moved, or the puppet was described as being heavy. A significant correlation between actual and simulated walking durations was observed only for the 7-year-olds in the informed condition.

The researchers concluded that the ability to imagine actions emerges in 7-year-olds when children are able to think about themselves in action using first-person egocentric based imagery. The researchers also noted that whereas research indicates that by 5 years children are able to use anticipatory mechanisms during goal-directed locomotion in a predictive way revealing a feed-forward control of their action (Grasso, Assaiante, Prévost, & Berthoz, 1998), 5-year-olds in their study were not able to explicitly imagine themselves acting. One might also speculate that the lack of correlation related to their lack of understanding of what the word "heavy" implied for action; therefore creating an action-language (embodied cognition) misrepresentation.

Caeyenberghs et al. (2009a) examined MI development in primary school children aged 7 – 12 years with the notion that such ability provides a window into the integrity of movement representation. MI ability was assessed using a pointing task and a mental hand rotation task, which was compared to motor skill ability. The relationship between MI and motor skill was shown to increase with age and most notably, the results indicated that significant improvement occurred after 7-8 years of age (representing the youngest group). Also in 2009, Frick and colleagues asked children 5- to 9 years of age and adults to tilt empty glasses, filled with varied amounts of imaginary water, so that the imagined water would reach the rim. In the manual tilting condition where glasses could be tilted 'actively' with visual feedback, even 5-year-olds performed well. However, in the imagined tilting condition performance for the 5-year-olds was poor. The researchers found that in order for the 5-year-olds to use MI, they had to engage self (motor), whereas the older children and adults were more reliant on visual information to solve the task. In conclusion, there was a clear age trend, indicating that bodily actions and motor feedback were particularly important in imagery performance of younger children.

Using an estimation of (perceived) reachability paradigm initially developed and reported on adults (e.g., Gabbard, Ammar, & Rodrigues, 2005; Gabbard, Ammar, & Lee, 2006; Gabbard, Cordova, & Lee, 2007a), Gabbard, Cordova, and Ammar (2007b) compared estimates of reach between children 5- to 11 years of age and young adults. Whereas there was no difference between groups for total error, a significant distinction emerged in reference to peripersonal [area within reach] and extrapersonal [area beyond reach] space. For children, significantly more error was exhibited with extrapersonal compared to peripersonal targets; there was no difference in adults. The groups did not differ in peripersonal space; however, adults were substantially more accurate in extrapersonal space. In addition, children revealed a greater overestimation bias. In essence, these data revealed a body-scaling problem in children for estimating reach in extrapersonal space; a difference that we initially hypothesized maybe due to developmental differences in use of visual information via egocentric (viewer dependent) and allocentric (viewer independent) representations. One could also speculate that the ability to mentally represent action may also be a factor. The researchers concluded that estimates of reachability in peripersonal space are adult-like as early as 5 years of age. However, the data also prompted speculation that the ability to map visual information from extrapersonal space for estimates of reach, emerge sometime between early adolescence (> 11 years and early adulthood). It should also be noted that the researchers pilot tested children as young as 3 years of age to determine ability to use the general experimental paradigm - they found that 5-year-olds were the most suitable population.

Citing that little has been done regarding the developmental nature of internal models (representations) of action, Choudhury et al. (2007a) measured the chronometry of executed and imagined hand actions of adolescents (mean age 13 years) and adults. For all participants, movement execution time significantly correlated with movement imagery time; however, there was a significant increase in the execution–imagery time correlation between adolescence and adulthood. Cognitive-motor efficiency per se did not change as indexed by both similar execution and imagery times on tasks for the adolescents and adults. Given that only the correlation between imagined and executed actions changed with age suggests that the developmental change was specific to generating accurate motor images and not a result of general cognitive improvement with age. Their results supported the notion that aspects of internal models are refined during adolescence. Furthermore, it was suggested that improvement was linked to development of the parietal cortex during adolescence; a point that is discussed in a subsequent section.

In a follow-up study, also using 13-year-olds, Choudhury et al. (2007b) reported similar results using a visually guided pointing motor task (VGPT) to test MI. Reaction time measures for both execution and imagery conditions in both adolescents and adults conformed to Fitts' Law. That is, accuracy decreased with a smaller target. However, the strength of association between conditions significantly increased with age, once again suggesting a refinement in action representation between adolescence and adulthood.

Brain Activity During MI. As noted earlier, Choudhury et al. (2007a, 2007b) refer to the link between development of the parietal cortex and action representation; suggesting that both are refined through the adolescent years. Supporting this idea are several other studies suggesting that the *parietal cortex* is involved in the formulation of internal models associated with MI and action representation (e.g., Blakemore, & Sirigu, 2003; Gerardin et al., 2000; Lacourse, Orr, Cramer, & Cohen, 2005).

Another aspect associated with the cognitive aspects of action representation, is development of the *prefrontal cortex*. That is, evolution of the transition of central cognitive mechanisms in terms of a general capacity to establish relations between data separated in space and time (Diamond, 2002). Molina et al. (2008) suggest that the evolution of MI in children can be interpreted in terms of a general development of cognitive processes involved in motor representation principally determined by internal changes in the prefrontal and parietal structures of the brain. In addition to the parietal and prefrontal cortices, research indicates that MI activates the supplementary motor area, the premotor and primary motor cortices, the basal ganglia and the cerebellum (e.g., Ehrsson, Geyer, & Naito, 2003; Fadiga & Craighero, 2004; see review by Munzert et al., 2009; Tomasino, Fink, Sparing, Dafotakis, & Weiss, 2008).

The Link to Embodied Cognition (EC)

Arguably, the notion of embodied cognition (EC) is one of the central issues in child development and learning. Whereas in recent years much has been reported on the subject, specifics regarding the link between cognitive processes and body action remain a mystery, especially regarding its developmental nature. Whereas several explanations regarding specific processes have been reported, the consensus of opinion for EC holds that cognitive processes are deeply grounded in our bodily interactions with the environment (e.g., Borghi & Cimatti, 2010; Barsalou, 2008; Bergen & Wheeler, 2010; Sadeghipour & Kopp, 2012). That is, cognitive representations and actions are inextricably linked – actions are central to the emergence of representations (Boncoddo, Dixon, & Kelley, 2010). Boncoddo et al. continue by suggesting, "if representations are grounded in action, then during the emergence of new representations, actions should play a critical role (p. 371)." One of the major propositions of this body of work is the notion that the motor system contributes to high-level cognitive processing. One of the first tactics used to study EC was the classic A-not-B error paradigm (see review by Thelen, Schöner, Scheier, & Smith, 2001). More recently, several innovative approaches have been used to study the connection between bodily actions and cognition in the context of, for example: language acquisition, numerical ability, and problem-solving skills (Anelli, Nicoletti, & Borghi, 2010; Boncoddo et al., 2010; Bergen & Wheeler, 2010; Domahs, Moeller, Huber, Willmes, & Nuerk, 2010; Rueschemeyer, Pfeiffer, & Bekkering, 2010).

Motor Cognition and Embodied Cognition. Another term frequently mentioned with studies of MI is *motor cognition*. This term commonly refers to the study of cognitive processes that drive action programming. MI is also considered a form of motor cognition (Gabbard, 2009). Motor cognition takes into account recognizing, anticipating, predicting and production of actions. The cognitive processes involved enable us not only to react to our environment but also to anticipate the consequences of our actions. Gallese, Rochat, Cossu, and Sinigaglia (2009) have proposed what they refer to as the motor cognition hypothesis. This idea, much like EC, suggests that the motor system plays a pivot role in cognitive functions. Furthermore, they suggest that the common neural mechanism shared with action and goal understanding is the mirror neuron system; a system frequently mentioned in associated with mental *simulation* (imagery) of action (e.g., Filimon, Nelson, Hagler, & Sereno, 2007; Molenberghs, Cunnington, & Mattingley, 2009; Munzert et al., 2009). Briefly, mirror neurons are cortical brain cells that become active when a particular behavior is performed and when that same behavior is observed; in essence, simulation, or perhaps more correct imitation, matches action in cognitive brain processes. Gallese and colleagues (2009) in their developmental review and hypothesis contend that a rudimentary level of the mirror neuron system is likely present at birth; a system that is flexible across time and influenced by motor experience and visuomotor learning. Although not mentioned specifically, the researchers make a strong case for embodied cognition by showing evidence that (for example) infants understand goal intentions in terms of their own motor knowledge. One of the key conclusions of their developmental treatise for a motor cognition hypothesis was that there is a motor account of the development of intentional (cognitive) understanding.

So, one might ask the question, is there a difference between motor cognition and EC? Obviously, the two are similar, but one might argue that it depends on the perspective of interest. That is, with motor cognition, emphasis is typically on 'motor' behavior and what cognitive structures and processes drive it. With EC, focus is commonly on how the environment and motor system influences cognitive behaviors. In any case, both share the role of simulation. For an excellent review of 'grounded motor cognition' in the form of gesture processing, refer to Sadeghipour and Kopp (2012).

The Role of Simulation in EC

A key finding in this literature that underscores the primary message of interest here is the idea that *simulation* plays a significant role in EC (e.g., Bergen & Wheeler, 2010; Brouillet, Heurley, Martin, & Brouillet, 2010; Engelen, Bouwmeester, Bruin, & Zwaan, 2011; Garbarini & Adenzato, 2004; Hostetter & Alibali, 2008). For example, Barsalou (2008) states, "Grounded cognition reflects the assumption that cognition is typically grounded in multiple ways, including *simulations*, situated action, and, on occasion, bodily states." Bergen and Wheeler state that when processing sentences, participants activate perceptual and motor systems to perform mental *simulations* of the events. Borghi and Cimatti present an excellent review of the role of *simulation* in EC with their personal explanation of what specifically is being simulated. Interestingly, as noted earlier, the authors link this aspect of EC with researchers associated with motor cognition and simulation in the form of MI (e.g., Decety & Grezes, 1999; Jeannerod, 2001). Adding to the general significance of simulated processes in information processing, Kosslyn, one of the foremost experts on the topic, declares that mental simulation underscores memory, reasoning, and learning (Kosslyn et al., 2006). For example, imagery processes rely in large part on retrieved episodic information – information used to generate explicit and accessible representations in working memory. The implications of such in regard to EC are remembered events that are observed through experience – events that are used to generate new representations. A review of the literature indicates that structural and functional brain circuitries supporting episodic memory experience significant reorganization during childhood (Shing et al., 2010).

The work of Boncoddo and colleagues (2010) provides an excellent example of simulation effects with children. The researchers asked preschoolers to solve a set of relatively simple gear-system problems. The researchers observed that children initially solved the problems by *simulating* the directional movements of the gears. That process led most participants to discover new representations of the problems. Results indicated that the number of actions that embodied alternation information during simulation predicted later emergence of higher-order representation. The researchers concluded that their findings were consistent with the EC hypothesis suggesting that actions are central to the emergence of new representations. It seems reasonable to conclude that the ability to simulate was a key factor in behavioral outcome.

Motor Imagery of Children with Atypical Development

Children with Developmental Coordination Disorder (DCD)

Citing that children with DCD have difficulties with MI and in generating an accurate visuospatial representation of an intended action, Maruff, Wilson, Trebilcock and Currie (1999) examined the chronometry of real and imagined movements using a visually-guided pointing task with a Fitts' law paradigm [speed for accuracy trade-offs that occur as target size varied for both real and imagined performance]. Performance of the control group conformed to Fitts' law. In the DCD group, only real movements conformed. The researchers concluded that children with DCD have an impairment in the ability to generate internal representations of volitional movements – an impairment known as an *internal modelling deficit (IMD) hypothesis.*

Wilson and colleagues (2004) further tested the *IMD hypothesis*, using a mental rotation paradigm. Participants included children with DCD and controls; both 10 years of age. The task required children to judge the handedness (side) of single-hand images that were presented at angles between 0° and 180° at 45° intervals in either direction. Responses of the control children conformed to the typical pattern of mental rotation: a moderate trade-off between response time and angle of rotation. However, the response pattern for the DCD group was less typical; that is, there was a small trade-off function. Response accuracy did not differ between groups. The researchers suggested that children with DCD, unlike controls, do not automatically engage MI when performing mental rotation, but rely on an alternative object-based strategy. It was concluded that these children manifest a reduced ability to make imagined transformations from an egocentric or first-person perspective.

Van Waelvelde and collegues (2006) also reported evidence indicating the children with DCD have problems with internal representation of action. The researchers used a rhythmic finger-tapping task in combination with a jumping and a drawing activity to evaluate children with DCD and a matched control group. In children with DCD, an increased temporal variability was found with finger tapping when compared with a control group. Errors in time correlated significantly with the jumping and drawing task, while errors in space did not. It was concluded that children with DCD have more problems in building up an internal representation of the movement. The researchers went on to suggest that deficits in temporal movement parameterization might be one of the underlying causes of poor motor performance in some children with DCD.

In 2006, Williams, Thomas, Maruff, Butson and Wilson also investigated the so-called *IMD* hypothesis using the mental rotation paradigm with typically developing and children with DCD (ages 7 to 11). Groups were asked to complete hand rotation tasks with and without explicit imagery instructions. In the instruction condition, participants were asked to imagine their own hand in the position of the stimulus. Overall, there were no significant differences between the DCD and control groups when the hand task was completed without explicit instructions, on either response time or accuracy. However, when imagery instructions were introduced, the controls were significantly more accurate than the DCD group, indicating that children with DCD were unable to benefit from explicit cuing. In addition to the hand tasks, participants also completed an alphanumeric task, and as predicted, no group difference was found. In summary, results indicated partial support for the *IMD* hypothesis with DCD children.

In a subsequent study (2008), Williams et al. examined MI ability in DCD children aged 7- to 11 years with the intent to determine whether children with varying degrees of motor impairment differed in their ability to perform MI tasks. DCD children were split into two groups (DCD severe [DCD-S], DCD mild [DCD-M]) and compared to age matched controls. Participants performed two MI tasks – hand (performed without and with specific imagery instructions) and whole-body rotation. Results indicated that children in the DCD-S group had a generalized MI deficit in that they were less accurate across tasks than controls (and the DCD-M group on the hand task) and showed little benefit when given specific imagery instructions. The DCD-M group appeared capable of performing simpler MI transformations, but was less successful as task complexity increased. Unlike the DCD-S group, the DCD-M group did show some benefit from specific imagery instructions with increases in accuracy on the hand task. The findings suggested that a MI deficit does exist in many children with DCD but that its degree can vary. That is, factors such as individual ability level and task complexity appear to be linked to the deficit.

Also in 2008, Lewis et al. examined movement durations for real and imagined movements in a visually guided pointing task in children aged 8- to 12 years old and (similar to previous studies noted) found that the DCD group demonstrated an inability to generate imagined movements that was not present in the typically developing participants. That same year, Deconinck, Spitaels, Fias, and Lenior (2008) investigated the *IMD* hypothesis with 9-year-old children with DCD using a mental (hand / letters) rotation paradigm using various stimulus orientations. Results indicated that children with DCD were generally slower and made more errors than the typically developing control sample. As expected, both groups displayed fewer errors and faster responses in the letter, compared to hand conditions. This complements the reasoning that the motor task requires additional resources compared to the visual (letter) task. The researchers concluded that both groups did use an imagery strategy; however, judgments of the DCD group seem to be compromised by a less well-defined internal model.

More recently, Williams, Omizzolo, Galea, and Vance (2012) investigated whether the MI deficits found in DCD children were linked to inattention, seen in about 50% of DCD children. The performance of four groups of children aged 7-12 years was compared using an imagined pointing task and a hand rotation task. Groups consisted of children with ADHD, ADHD + DCD, DCD alone, and a control. With the imagined pointing task, children with

DCD did not conform to speed accuracy trade-offs during imagined movements, but all other groups did. However, on the hand rotation task, both the ADHD + DCD and DCD groups were less accurate than the ADHD and control groups. This study demonstrated that children with ADHD + DCD experience genuine motor control impairments manifested by a reduced ability to accurately represent movement – a problem that does not appear to be linked to increased levels of inattention or decreased working memory capacity typically observed in children with ADHD alone.

In summary, the *IMD* hypothesis suggests that one of the underlying causes of DCD may be an inability to utilize internal models of motor control accurately. Converging support for the *IMD* hypothesis suggests quite convincingly that there is strong relationship with the child's inability to mentally represent action in the form of MI.

Possible Brain Impairments Associated with DCD

The literature suggests that the root of DCD could be in several cortical areas involved in motor processing, such as the premotor area (plays a role in motor planning and sequencing and movement readiness) and areas of the parietal lobes and the dorsolateral prefrontal cortex, which are also associated with premotor planning (Ehrsson et al., 2003; Munzert et al., 2009; Pelgrims, Andres, & Olivier, 2009). Another brain structure, not mentioned in previous work with typically developing children, is the cerebellum. Caeyenberghs and colleagues (2009b) suggested that the cerebellum may play an important role in the reduction of MI accuracy, since it is thought to encode internal models that reproduce the essential properties of action representations. Most researchers of DCD agree that the deficits displayed by these children are the result of a general processing deficit of the motor system as a whole (Piek & Pitcher, 2004). For example, when information "flows" from the prefrontal to the motor cortex, DCD children, especially younger ones, may be forced to maintain a high level of pre-programming to compensate for difficulties in the perceptual-motor and executive aspects of the movement related to their coordination disorder (Castelnau, Albaret, Chaix, & Zanone, 2008). Lewis et al. (2008) have suggested that the inability of children with DCD to motor image is caused by some disruption to normal functioning of the inferior parietal lobe (part of the parietal cortex); as noted earlier, this area of the brain is thought to be a crucial site for activating covert motor processes during simulated actions; this observation was confirmed in recent review of the brain structure by (Creem-Regehr, 2009). The researchers went on to propose that with DCD there is an impairment to constructing forward models. In addition, the original motor intention appears to emanate from the premotor cortex; however, a functional loop between premotor and parietal cortex appears to subserve this feedforward planning process (Sirigu et al., 2004). This process provides stability to the motor system by *predicting* and anticipating the sensory *consequences* of self-generated movements (Grush, 2004; Wolpert, 1997; Pelgrims, Andres, & Olivier, 2005). Thus, even if the movement is initially executed correctly, the normal ability to correct the movement pathway during the action may not be available (Ameratunga, Johnston, & Burns, 2004). At this point, the primary question is: is the delay displayed in children with DCD different from the developmental immaturity of typically developing children?

Children with Other Motor Disorders

MI has also been explored in children with cerebral palsy (CP) and spastic hemiplegia (SH). Crajé, Aarts, der Sanden, and Steenbergen (2010) investigated action planning in young children with unilateral CP and a control group (ages 3 – 6 years). Participants performed a sequential movement task and were tested for end-state comfort effect; asking, "do participants adapt their initial grip choice so they would have a comfortable posture at the end of the action?" The experimental setup examines evidence for anticipatory planning. The results showed that planning was impaired and there was no age effect in the unilateral CP group when compared to the control group. However, a specific 8-week intervention program improved anticipatory planning in children with CP, confirming that the use of MI may be a promising technique to train motor planning in children with unilateral CP.

MI deficits have also been found in children with spastic hemiplegia (SH), a condition characterized by muscle spasticity on one side of the body that affects motor skill execution. SH is also considered a form of CP, if congenital. Williams, Reid, Reddihough, and Anderson (2011) compared performance in a hand rotation task in children with left and right hemiplegia to a control group. Even though there was a general slowing of responses in the hemiplegia groups, the difference from the typically developing group was not significant. Performance in the SH groups was linked to functional level, reinforcing the notion that MI processes depend highly on motor activity. The same research group also compared performance of 8- to 12-year-old children with SH to children with DCD using rotation tasks involving hand and whole body (Williams, Anderson, Reddihough, Reid, Vijayakumar, & Wilson, 2011). Results indicated that both groups were impaired in their ability to perform the tasks as accurately as their typically developing peers. The researchers concluded that the reduced ability to utilize MI was part of an underlying motor planning deficit in both groups, however, it remains unclear whether this deficit was the result of impaired neural networks (that cause the condition) or due to motor execution difficulties (consequence of the condition).

Conclusion

From the research described here involving typically developing children, one could conclude that the emergence of MI begins around age 5 (Funk, Brugger, & Wilkening, 2005; Gabbard et al., 2007b; Kosslyn et al., 1990) with considerable refinement in ability shown between early adolescence (12 yrs >) and early adulthood (Choudhury et al., 2007a, 2007b; Gabbard et al., 2007b; Gabbard, Cordova, & Lee, 2009). As a general observation of studies involving a range of age, older populations perform better than younger groups (e.g., Caeyenberghs et al., 2009a; Molina et al., 2008).). In other words, the relationship between MI and motor skills gets stronger with increasing age. Furthermore, those findings suggest that the ability to mentally represent actions (i.e., action representations) emerges around the same time.

Most studies of special populations indicate that MI ability seems to be dependent on the level of motor skill allowed by the disorder. For example, young children with CP show impaired action planning when compared to typically developing children. Studies of children with DCD (ages 7- to 12 years) indicate that the motor skill deficits observed in the condition are associated with poor performance in MI tasks. Arguably, the importance of that body of research is in large part the insight that the findings have provided regarding brain activity and workings of the internal model structure; information that may be used to address typically developing children and those with other motor disorders.

The use of MI as an experimental paradigm for the study of cognitive aspects of action planning and control is a widely acceptable tactic with adult populations. With the intent to grasp an understanding of the developmental course of action representation and planning, studies are emerging with focus on younger populations. The fruits of this line of inquiry are promising with typically developing children and with those children that have shown motor deficits (e.g., DCD, SH, CP, etc). Results of current studies establish that through the use of MI, it is possible to assess movement opportunities and foresee consequences before performing a motor skill.

More important, is the notion that the assessment of MI integrity may help educators and therapists diagnose and treat movement disorders. MI training offers clinicians and therapists the opportunity to directly stimulate cortical areas that otherwise cannot be activated, such as the cortical area representing a paralyzed limb. MI can strengthen intact neural pathways and facilitate activation of motor cortical areas. Multiple studies have demonstrated that action observation and MI can elicit sensorimotor cortical activity and, at least temporarily, improve motor functions of paralyzed limbs after stroke and spinal cord injury (Wang et al., 2011).

MI training can result in improvements in speed, accuracy, and strength of motor execution (Lorey et al., 2010). Obviously, before considering using MI training, it is important to determine whether a child can perform MI (especially when the child displays signs of perceptual-motor disorders) and what tasks are more suitable for the age of the child. Finally, it is important to consider how the MI task interacts with and facilitates other activities that are included in the training / rehabilitation protocol.

And finally, we conclude that the concept of MI has established itself as a fundamental topic of study in the field of motor behavior. MI can be used for training and improvement of movement skills, rehabilitation, and also to assess the integrity of the motor system. Obviously, mechanisms of MI depend on the nature of the task, the level of skill, and individual capacity (Takahashi et al., 2005). For research and application purposes, MI may be used as a tool for intervention and / or rehabilitation with healthy and clinical populations, therefore providing helpful insights into motor behavior mechanisms. As a framework for training and rehabilitation in children, MI can be equally effective to perceptual-motor training.

References

Ameratunga, D., Johnston, L., & Burns, Y. (2004). Goal-oriented upper limb movements by Children with and without DCD: A window into perceptuo-motor dysfunction? *Physiotherapy Research International, 9*(1), 1-12.

Anelli, F., Nicoletti, R., & Borghi, A. M. (2010). Categorization and action: What about object consistence? *Acta Psychologica, 133*(2), 203-211.

Barsalou L.W. (2008). Grounded cognition. *Annual Review of Psychology, 59*, 617-645.

Bergen, B., & Wheeler, K. (2010). Grammatical aspect and mental simulation. *Brain and Language, 112*(3), 150-158.

Blakemore, S. J., & Sirigu, A. (2003). Action prediction in the cerebellum and in the parietal lobe. *Experimental Brain Research, 153,* 239-245.

Boncoddo, R., Dixon, J. A., & Kelley, E. (2010). The emergence of novel representations from action: evidence from preschoolers. *Developmental Science, 13*(2), 370-377.

Borghi, A. M., & Cimatti, F. (2010). Embodied cognition and beyond: Acting and sensing the body. *Neuropsychologia, 48*(3), 763-773.

Bourgeois, J., & Coello, Y. (2009). Role of inertial properties of the upper limb on the perception of the boundary of personal space. *Psychologie Française*, *54*(3), 225-239.

Brouillet, T., Heurley, L., Martin, S., & Brouillet, D. (2010). The embodied cognition theory and the motor component of "yes" and "no" verbal responses. *Acta Psychologica, 134*(3), 310-317.

Caeyenberghs, K., Tsoupas, J., Wilson, P. H., & Smits-Engelsman, B. C. M. (2009). Motor imagery in primary school children. *Developmental Neuropsychology, 34*(1), 103– 121.

Caeyenberghs, K., van Roon, D., Swinnen, S.P., & Smits-Engelsman, B.C.M. (2009). Deficits in executed and imagined aiming performance in brain-injured children. *Brain and Cognition*, *69*(1), 154-161.

Castelnau, P., Albaret, J., Chaix, Y., & Zanone, P. (2008). A study of EEG coherence in DCD children during motor synchronization task. *Human Movement Science*, *27*(2), 230-241.

Choudhury, S., Charman, T., Bird, V., & Blakemore, S. (2007a). Development of action representation during adolescence. *Neuropsychologia*, *45*, 255-262.

Choudhury, S., Charman, T., Bird, V., & Blakemore, S. (2007b). Adolescent development of motor imagery in a visually guided pointing task. *Consciousness and Cognition, 16(4),* 886-896.

Csibra, G., Bírób, S., Koósc , O., & Gergely, G. (2003). One-year-old infants use teleological representations of actions productively. *Cognitive Science, 2*(1), 111-133.

Crajé, C., Aarts, P., Nijhuis-van der Sanden, M., & Steenbergen, B. (2010). Action planning in typically and atypically developing children (unilateral cerebral palsy). *Research in Developmental Disabilities, 31,* 1039-46.

Creem-Regehr, S. H. (2009). Sensory-motor and cognitive functions of the human parietal cortex involved in manual actions. *Neurobiology of Learning and Memory, 91*(2), 166-171.

Decety, J., & Grezes, J. (1999). Neural mechanisms subserving the perception of human actions. *Trends in Cognitive Sciences, 3*, 172-178.

Deconinck, F. J., Spitaels, L., Fias, W., & Lenior, M. (2008). Is developmental coordination disorder a motor imagery deficit? *Journal of Clinical and Experimental Neuropsychology, 1*, 1-11.

Diamond, A. (2002). Normal development of prefrontal cortex from birth to young adulthood:

Cognitive functions, anatomy, and biochemistry. In: D.T. Stuss & R.T. Knight (Eds.), *Principles of frontal lobe function*, Oxford University pp. 466–503.

Domahs, F., Moeller, K., Huber, S., Willmes, K., & Nuerk, H. (2010). Embodied numerosity: Implicit hand-based representations influence symbolic number processing across cultures. *Cognition, 116*, 251-266.

Engelen, J. A. A., Bouwmeester, S., Bruin, A. B. H., & Zwaan, R. A. (2011). Perceptual simulation in developing language comprehension. *Journal of Experimental Child Psychology, 110*, 659-775.

Ehrsson, H. H., Geyer, S., & Naito, E. (2003). Imagery of voluntary movements of fingers, toes, and tongue activates corresponding body-part specific motor representations. *Journal of Neurophysiology, 90,* 304–3316.

Fadiga, L., & Craighero, L. (2004). Electrophysiology of action representation. *Journal of Clinical Neurophysiology, 21*, 157–169.

Filimon, F., Nelson, J. D., Hagler, D. J., & Sereno, M. I. (2007). Human cortical representations for reaching: Mirror neurons for execution, observation, and imagery. *NeuroImage, 37*(4), 1315-1328.

Frick, A., Daum, M. M., Wilson, M., & Wilkening, F. (2009). Effects of action on children's and adults' mental imagery. *Journal of Experimental Child Psychology, 104*, 34-51.

Funk, M., Brugger, P., & Wilkening, F. (2005). Motor processes in children's imagery: The case of mental rotation of hands. *Developmental Science, 8* (5), 402-408.

Gabbard, C. (2009). Studying action representation in children via motor imagery. *Brain and Cognition, 71*(3), 234-239.

Gabbard, C., Ammar, D., & Lee, S. (2006). Perceived reachability in single- and multiple degree of freedom workspace, *Journal of Motor Behavior, 38*(6), 423-430.

Gabbard, C., Ammar, D., & Rodrigues, L. (2005). Perceived reachability in hemispace, *Brain and Cognition, 58*(2), 172-177.

Gabbard, C., Cordova, A., & Ammar, D. (2007b). Children's estimation of reach in peripersonal and extrapersonal space, *Developmental Neuropsychology, 32*(3), 749- 756.

Gabbard, C., Cordova, A., & Lee, S. (2007a). Examining the effects of postural constraints on estimating reach. *Journal of Motor Behavior, 39*(4), 242-246.

Gabbard, C., Cordova, A., & Lee, S. (2009). Do children perceive postural constraints when estimating reach (motor planning)? *Journal of Motor Behavior, 41*(2), 100–105.

Gallese, V., Rochat, M., Cossu, G., & Sinigaglia, C. (2009). Motor cognition and its role in the phylogeny and ontogeny of action understanding. *Developmental Psychology, 45*(1), 103-113.

Garbarini, F., & Adenzato, M. (2004). At the root of embodied cognition: Cognitive science meets neurophysiology. *Brain and Cognition, 56*(1), 100-106.

Gerardin, E., Sirigu, S., Lehericy, A., Poline, J.B., Gaymard, B., Marsault,C., Agid, Y., & Le Bihan, D. (2000). Partially overlapping neural networks for real and imagined hand movements. *Cerebral Cortex, 10*, 1093–1104.

Grasso, R., Assaiante, C., Prévost, P., & Berthoz, A. (1998). Development of anticipatory orienting strategies during locomotor tasks in children, *Neuroscience & Biobehavioral Reviews, 22*, 533–539.

Grush, R. (2004). The emulation theory of representation: motor control, imagery, and perception. *Behavioral and Brain Sciences, 27*(3), 377-396.

Hostetter, A. B., & Alibali, M. W. (2008). Visible embodiment: gestures as simulated action. *Psychonomic Bulletin & Review, 15*, 495–514.

Jeannerod, M. (2001). Neural simulation of action: a unifying mechanism for motor cognition. *Neuroimage, 14*, 103-109.

Kosslyn, S. M., Margolis, J. A., Barrett, A. M., Goldknopt, E., & Daly, P. F. (1990). Age differences in imagery abilities. *Child Development, 61*, 995-1010.

Kosslyn, S. M., Thompson, W. L., & Ganis, G. (2006). *The case for mental imagery*. New York: Oxford University Press.

Kunz, B. R., Creem-Regehr, S. H., & Thompson, W.B. (2009). Evidence for motor simulation in imagined locomotion. *Journal of Experimental Psychology: Human Perception and Performance, 35*(5), 1458-1471.

Lacourse, M. G., Orr, E.L., Cramer, S.C., & Cohen, M.J. (2005). Brain activation during execution and motor imagery of novel and skilled sequential hand movements, *Neuroimage, 27*, 505–519.

Lewis, M., Vance, A., Maruff, P., Wilson, P., & Cairney, S. (2008). Differences in motor imagery between children with developmentl coordination disorder with and without the combined type of ADHD. *Developmental Medicine & Child Neurology, 50*(8), 608-612.

Lorey, B., Pilgramm, S., Walter, B., Stark, R., Munzert, J., & Zentgraf, K. (2010). Your Mind's eye: Motor imagery of pointing movements with different accuracy. *Neuroimage, 49*(4), 3239-3247.

Maruff, P., Wilson, P., Trebilcock, M., & Currie, J. (1999) Abnormalities of imagined motor sequences in children with developmental coordination disorder. *Neuropsychologia, 37*, 1317-1324.

Molenberghs, P., Cunnington, R., & Mattingley, J. B. (2009). Is the mirror neuron system involved in imitation? A short review and meta-analysis. *Neuroscience & Biobehavioral Reviews, 33*(7), 975-980.

Molina, M., Tijus, C., & Jouen, F. (2008). The emergence of motor imagery in children. *Journal of Experimental Child Psychology*, *99*(3), 196-209.

Moulton, S. T., & Kosslyn, S. M. (2009). Imagining predictions: mental imagery as mental emulation. *Philosophical Transactions of the Royal Society, 364*, 1273-1280.

Munzert, J., Lorey, B., & Zentgraf, K. (2009). Cognitive motor processes: The role of motor imagery in the study of motor representations. *Brain Research Reviews*, *60*(2), 306-326.

Pelgrims, B., Andres, M., & Olivier, E. (2005). Motor imagery while judging object-hand interactions. *NeuroReport, 16*(11), 1193-1196.

Pelgrims, B., Andres, M., & Olivier, E. (2009). Double dissociation between motor and visual imagery in the posterior parietal cortex. *Cerebral Cortex, 19*(10), 2298-2307.

Piek, J., & Pitcher, T. A. (2004). Processing deficits in children with movement and attention deficits. In D. Dewey & D. E. Tupper (Eds.). *Developmental motor disorders*. New York: Guilford Press.

Rueschemeyer, S., Pfeiffer, C., & Bekkering, H. (2010). Body schematics: On the role of the body schema in embodied lexical-sematic representations. *Neuropsychologia, 48*(3), 774-781.

Sadeghipour, A., & Kopp, S. (2012). Gesture processing as grounded motor cognition: Towards a computational model. *Procedia – Social and Behavioral Sciences*, *32*, 213-223.

Schubotz, R. I. (2007). Prediction of external events with our motor system: Towards a new framework. *Trends in Cognitive Sciences, 11*(5), 211-218.

Sharma, N., Jones, P. S., Carpenter, T. A., & Baron, J. (2008). Mapping the involvement of BA 4a and 4p during motor imagery. *NeuroImage, 41*(1), 92-99.

Sharma, N., Simmons, L. H., Jones, P. S., Day, D. J., Carpenter, T. A., Pomeroy, V. M., Warburton, E. A., & Baron, J. (2009). Motor imagery after subcortical stroke: A functional magnetic resonance imaging study. *Stroke, 40*, 1315 - 1324.

Shing, Y. L., Werkle-Bergner, M., Brehmer, Y., Muller, V., Li, S. K., & Lindenberger, U. (2010). Episodic memory across the lifespan: The contributions of associative and strategic components. *Neuroscience & Biobehavioral Reviews*, *34*(7), 1080-1091.

Sirigu, A., Daprati, E., Ciancia, S., Giraux, P., Nighoghossian, N., Posada, A., & Haggard, P. (2004). Altered awareness of voluntary action after damage to the parietal cortex. *Nature Neuroscience, 7*(1), 80-84.

Skoura, X., Vinter, A., & Papaxanthis, C. (2009). Mentally simulated motor actions in children. *Developmental Neuropsychology, 34*, 356-367.

Smits-Engelsman, B. C. M., & Wilson, P. (in press, 2012). Age-related changes in motor imagery from early childhood to adulthood: Probing the internal representation of speed-accuracy trade-offs. *Human Movement Science*, on-line first.

Solodkin, A., Hlustik, P., Chen, E. E., & Small, S. L. (2004). Fine modulation in network activation during motor execution and motor imagery. *Cerebral Cortex, 14*, 1246 - 1255.

Springer, A., Hamilton, A. F. C., & Cross, E. S. (2012). Simulating and predicting others' actions. *Psychological Research, 76*(4), 383-387.

Sommerville, J. A., Woodward, A. L., & Needham, A. (2005). Action experience alters 3-month-old infants' next term perception of others' actions. *Cognition, 96*(1), B1-B11.

Suddendorf, T., & Moore, C. (2011). Introduction to the special issue: The development of espisodic forsight. *Cognitive Development, 26*, 295-298.

Stanford Encyclopedia of Philosophy. (2008). *Mental representation*. Open access on-line Encyclopedia.

Stevens, J. A. (2005). Interference effects demonstrate distinct roles for visual and motor imagery during the mental representation of human action. *Cognition, 95*(3), 329-350.

Takahashi, M., Hayashi, S., Ni, Z., Yahagi, S., Favilla, M., & Kasai, T. (2005). Physical practice induces excitability changes in human hand motor area during motor imagery. *Experimental Brain Research, 163*(1), 132-136.

Thelen, E, Schoner, G, Scheier, C, & Smith, L. B. (2001). The dynamics of embodiment: A field theory of infant perseverative reaching. *Behavioral and Brain Sciences*, *24*(1), 1-86.

Tomasino, B., Fink, G. R., Sparing, R., Dafotakis, M., & Weiss, P. H. (2008). Action verbs and the primary motor cortex: A comparative study of silent reading, frequency judgments, and motor imagery. *Neuropsychologia, 46*(7), 1915-1926.

van Waelvelde, H., de Weerdt, W., de Cock, P., Janssens, L., Feys, H., Bouwien, C. M., & Smits-Engelsman, B. C. (2006). Parameterization of movement execution in children with developmental coordination disorder. *Brain and Cognition, 60*(1), 20-31.

Wang, W., Collinger, J. L., Perez, M., Tyler-Kabara, E., Cohen, L., Birbaumer, N., Brose, S., Schwartz, A., Boninger, M., & Weber, D. (2011). Neural interface technology for rehabilitation: Exploiting and promoting neuroplasticity. *Physical Medicine and Rehabilitation Clinics of North American*, *21*(1), 157–178.

Willatts, P. (1999). Development of means-end behavior in young infants: Pulling a support to retrieve a distant object. *Developmental Psychology 35*, 651–667.

Williams, J., Thomas, P. R., Maruff, P., Butson, M. & Wilson, P. H. (2006). Motor, visual and egocentric transformations in children with Developmental Coordination Disorder. *Child: Care, health and development*, *32*, 633–647.

Williams, J., Thomas, P. R., Maruff, P., & Wilson, P. H. (2008). The link between motor Impairment level and motor imagery ability in developmental coordination disorder. *Human Movement Science*, *27*(2), 270-285.

Williams, J., Omizzolo, C., Galea, M. P., & Vance, A. (2012). Motor imagery skills of children with attention hyperactivity disorder and developmental coordination disorder. *Human Movement Science,* online first.

Williams, J., Reid, S. M., Reddihough, D. S., & Anderson, V. (2011). Motor imagery ability in children with congenital hemiplegia: Effect of lesion side and functional level. *Research in Developmental Disabilities, 32,* 740-748.

Williams, J., Anderson, V., Reddihough, D. S., Reid, S. M., Vijayakumar, N., & Wilson, P. (2011). A comparison of motor imagery performance in children with spastic hemiplegia and developmental coordination disorder. *Journal of Clinical and Experimental Neuropsychology, 33*(3), 273-282.

Wilson, P. H., Maruff, P., Butson, P., Williams, J., Lum, J., & Thomas, P. R. (2004). Internal representation of movement in children with developmental coordination disorder: A mental rotation task. *Developmental Medicine & Child Neurology, 46,* 754-759.

Wolpert, D. M. (1997). Computational approaches to motor control. *Trends in Cognitive Sciences, 1,* 209-216.

Young, S. J., Pratt, J, & Chau. T. (2009). Misperceiving the speed-accuracy tradeoff: Imagined Movements and perceptual decisions. *Experimental Brain Research, 192*(1), 121-132.

In: Motor Behavior and Control: New Research
Editors: Marco Leitner and Manuel Fuchs
ISBN: 978-1-62808-142-8

Chapter 3

Cognitive Training Enhances Motor Performance and Learning

John S. Y. Chan[1], *Jin H. Yan*[2,3] *and V. Gregory Payne*[4]
[1]Department of Psychology, The Chinese University of Hong Kong, Hong Kong SAR, China
[2]Sichuan Research Center of Applied Psychology, Chengdu Medical College, Chengdu, China
[3]Department of Psychology, Tsinghua University, Beijing, China
[4]Department of Kinesiology, San Jose State University, San Jose, CA, US

Abstract

Cognition and motor performance are interdependent for optimal daily functioning. In recent years, the role of cognitive training for able-bodied and disabled populations has been widely investigated. The importance of cognitive training for cognitive and motor improvement has generally received strong empirical support. There is evidence that non-trained domains can also be enhanced (*e.g.*, sport-specific creativity and driving). The findings suggest that motor performance and learning can be improved when viable cognitive training paradigms are used. In this chapter, we first examine the roles of attention, memory, and imagery on motor performance and learning. Second, we discuss how cognitive training can enhance motor abilities. Third, the feasibility and prospect of computerized cognitive training on motor performance and learning is considered. Finally, practical implications related to cognitive training, motor performance, and rehabilitation are examined as related to the need for future research.

Introduction

Motor behavior, motor skills, and cognition are interdependent and necessary for optimal human functioning. Development in both motor and cognitive domains generates similar trajectories across the human lifespan (Ren, Wu, Chan, and Yan, 2013; Yan, Thomas, and

Payne, 2002; Yan, Thomas, Stelmach, and Thomas, 2000). In general, most cognitive and motor abilities improve markedly in childhood, but show signs of deterioration as one approaches old age (Yan, 2000). The relationship between cognitive and motor capabilities leads to the hypothesis that motor enhancements can be achieved through cognitive improvements. This topic will be discussed in this chapter. Herein, motor capability refers to a combination of motor performance (motor proficiency at a given time) and motor learning (improvement of motor proficiency over time). Furthermore, for our purposes in this chapter, cognitive training is a program designed to enhance a person's mental skills, such as attention, memory and imagery; whereas physical training or practice is to enhance motor skill through personal and direct physical experience of the skill without additional mental assistance.

Enhancing motor performance and motor learning has traditionally involved direct physical or motor skill practice (*e.g.* for learning ball juggling, a person actually juggles balls and learns the skills through trial and error). However, new ways to improve motor performance have emerged. As reported in a recent meta-analytic study, expert athletes outperform novice performers in processing speed and visual attention (Voss, Kramer, Basak, Prakash, and Roberts, 2010). Perhaps, enhancing one's cognitive performance, in addition to traditional physical or motor skill training, may contribute to one's motor proficiency. In recent years, different cognitive training paradigms have been devised, and the viability of cognitive training for motor improvement has been more strongly considered. This is supported by a growing body of research (*e.g.* Hagemann, Strauss, and Canal-Bruland, 2006; Moore, Vine, Cooke, Ring, and Wilson, 2012; Post, Muncie, and Simpson, 2012). For example, cricketers have benefited from visual-perceptual training to improve their decision making accuracy in computer and field tests (Hopwood, Mann, Farrow, and Nielsen, 2011). Thus, cognitive training can reasonably be assumed to improve motor performance and learning when applied alone or supplemental to traditional physical training.

In this chapter, three important cognitive abilities (attention, memory, and imagery) will be explored. Their contributions to motor skill learning and motor development will also be explicitly described with rationales related to motor improvement through cognitive training being provided. Moreover, the feasibility of computerized cognitive training to augment motor performance and learning will be discussed. And lastly, some practical considerations and future research directions will be addressed.

Cognitive Abilities and Motor Skills

Attention

Attention refers to an allocation of cognitive resources to a specific object or the specific aspect of the object. Attention helps us glean and discern critical information from our environment. In a state of reduced attention, or being distracted by irrelevant stimuli, the ability to process relevant information and make corresponding and appropriate motor adjustments is impaired (Kam *et al.*, 2012).

Motor performance requires considerable attentiveness, even for actions as common and seemingly simple as walking (Abbud, Li, and DeMont, 2009). The ability to sustain attention while walking is related to walking velocity (Hotltzer, Wang, and Verghese, 2012). In addition, attentional processes and movement planning are related (Ridderikhoff, Peper, and Beek, 2008). For dancers, attention helps timing control, especially for complicated dance sequences and choreography (Minvielle-Concla, Audiffren, Macar, and Vallet, 2008). In sports, athletes of different levels of expertise show varying attentional performance; hence, a possible solution to enhancing one's dance and sport performance could be attentional training. As reported in a meta-analysis, expert athletes generally can ascertain more information from fewer, but longer, fixations than those less skillful (Mann, Williams, Ward, and Janelle, 2007). Compared to non-athletes, orienteers are better able to direct their attention to the center or periphery of a visual field (Cereatti, Rasella, Manganelli, and Pesce, 2009). More experienced orienteers can attend more to a map while moving. This is rarely observed among those with less experience (Eccles, Walsh, and Ingledew, 2006). Experienced basketball players are less vulnerable to inattentional blindness (a tendency to ignore information of the unattended parts) than non-experienced players (Furley, Memmert, and Heller, 2010), and expert handball players with higher scores in attention measures tend to be more accurate than those with lower scores in sport-specific thinking abilities (Memmert, 2011). Athletes from many sports perform better with greater attention abilities. In addition, some sports require more attention than others to be successful. As attentional demands of a task increase with target speed (Holcombe and Chen, 2012), higher attentional ability may become increasingly important for players of fast moving sports (*e.g.* table tennis and baseball).

In addition to motor performance, attention also interacts with motor learning. When one's attention is divided or distracted, motor skill learning can be impaired (Taylor and Thoroughman, 2007). Some training programs emphasizing attentional efforts have been designed and tested, with varying results. A common method to train attention is "cueing" where a cue (*e.g.* color patch) is used to direct the trainee's attention to the critical information (*e.g.* opponent's shoulder movement). This has been effective in improving anticipation skills in novice and local league badminton players, but not in national league level players (Hagemann, Strauss, and Canal-Bruland, 2006). Such improvements have not been observed among handball goalkeepers (Abernethy, Schorer, Jackson, and Hagemann, 2012). The type of sport and the level of expertise seem to moderate the cueing effectiveness of the training program.

Different from the cueing method, the quiet eye paradigm teaches the trainee to have longer final fixation on the target before movement execution. Quiet eye training increases basketball shooting accuracy under high pressure (Vine and Wilson, 2011). Compared with golfers receiving technical training, quiet eye trainees exhibit higher putting accuracy and more effective gaze control in retention testing or testing under pressure (Moore, Vine, Cooke, Ring, and Wilson, 2012). Moreover, some researchers have tried to broaden one's attention span so as to enhance motor performance. Attention-broadening programs have resulted in better sport-specific creativity, especially in complex tasks related to team sports in real-world situations (Memmert, 2007).

Over the past decade, considerable research related to attentional focus has been conducted. Much of this research provides evidence and insights into the development of cognitive and motor training. One general finding is that emphasis on external focus of attention (focus on movement effects) is superior to internal focus of attention (focus on movement itself) in motor performance and acquisition of motor skills. This has been demonstrated in many sports and activities including skiing, golfing, balancing, force production, and long jump (Lohse, Sherwood, and Healy, 2011; Porter, Anton, and Wu, 2012; Wulf, Höß, and Prinz, 1998; Wulf, Lauterbach, and Toole, 1999; Wulf, Shea, and Park, 2001). An external focus of attention can enhance movement economy in running and dart throwing, promote utilization of more natural movement mechanisms and increase the learner's resistance to external distractions (Lohse, Sherwood, and Healy, 2010; McNevin, Shea, and Wulf, 2003; Poolton, Maxwell, Masters, and Raab, 2006; Schucker, Hagemann, Strass, and Volker, 2009).

External focus of attention can benefit players at different levels of expertise (Wulf and Su, 2007). It can improve learner's performance in both practice and transfer sessions (Totsika and Wulf, 2003). Although external focus of attention is beneficial to motor learning in general, trainees' ages or developmental states should be considered in various training paradigms. Comparable benefits have been attained in older and young adults (Chiviacowsky, Wulf, and Wally, 2010). However, for children, the benefits are more subtle. For example, research involving dart throwing learning indicated that adults' retention and transfer performances benefit from external attentional focus while children's transfer performance benefits from internal focus of attention (Emanuel, Jarus, and Bart, 2008). Perhaps, external and internal foci of attention are more suitable for adults and children, respectively, during motor learning. Though learning of a variety of motor skills can be improved with external focus of attention, it is not as effective for learning sports that emphasize form (*e.g.* gymnastics routine) (Lawrence, Gottwald, Hardy, and Khan, 2011). Thus, some caution is warranted in broadly providing attention assessment and training to athletes. Previous research has suggested that there is no difference in basic attentional measures among athletes from different expertise levels (Memmert, Simons, and Grimme, 2009). Therefore, to assess athletes' attention, sport-specific measures are more appropriate, and sport-specific trainings are more likely to lead to performance improvements in that sport.

Memory

Efficient memory helps learning and performance of motor skills. Retrieval of task-related motor representations and the consolidation of short term memory take place when a motor skill is executed and learned. Previous research generally suggests that motor learning is dependent on working memory. Memory consolidation is relevant to expert performance (Furley and Memmert, 2012; Maxwell, Masters, and Eves, 2003; Schack and Mechsner, 2006).

Expertise and memory performance are related. Elite basketball players can retain memory better than novices (Kioumourtzoglou, Derri, Tzetzis, and Theodorakis, 1998). Expert snooker players can recall and recognize ball patterns more accurately than novices, though they have similar vision (Abernethy, Neal, and Koning, 1994). Expert soccer players have more elaborate memory organization than novices to facilitate encoding and storing visual patterns in abstract representation in memory (Poplu, Ripoli, Mavromatis, and Baratgin, 2008; Williams, Davis, Burwitz, and Williams, 1993). Furthermore, compared to less experienced players, expert soccer players need less time to recognize familiar and unfamiliar soccer action sequences (Williams, Hodges, North, and Barton, 2006). Besides motor expertise, age is also a contributing factor as it relates to memory and motor performance in children, perhaps due to the development of a more flexible neural system (Ward and Williams, 2003). In addition to long term memory, short term memory or working memory (a temporary store to sustain information that is readily retrieved and manipulated) is also linked to motor performance. Athletes with high working memory capacity have been shown to control attention more efficiently in complex situations. This is especially useful in decision making in sport settings (Furley and Memmert, 2012). A neuroimaging study indicates that during the performance, expert archers recruit more cortical areas for spatial working memory than their novice counterparts (Seo *et al.*, 2012).

However, there is ongoing debate over the issue of domain specificity of expert memory. For instance, questions remain about whether expert memory in one sport can be similarly observed in other sports or non-sport settings. Memory expertise is usually domain-specific. For example, experienced basketball players show similar spatial memory capacity to that of college students who are not team-ball participants (Furley and Memmert, 2010). This suggests that non-domain-specific memory measures can not differentiate basketball players from non-players. However, this does not mean that memory expertise has to be domain-specific; the situation is far more intriguing and complex. As observed in another study, expert memory is not entirely domain-specific. Experts are better at remembering patterns similar to what they encounter in their field of expertise. Expert players of different sports are better able than non-expert players to recall positions of defensive players in sports with which they are unfamiliar (Abernethy, Baker, and Côté, 2005).

In addition, working memory plays an important role in motor learning. Information to be learned is stored in the working memory before it is consolidated into long term memory. Hence, in motor learning, the magnitude of one's working memory capacity is important in the early stage of skill acquisition when a learner has a relatively low skill proficiency (Anguera *et al.*, 2012; Anguera, Reuter-Lorenz, Willingham, and Seidler, 2010; Maxwell, Masters, and Eves, 2003). In young adults, spatial working memory performance is significantly correlated with motor learning. The neural correlates of both working memory and motor learning are quite similar (Langan and Seidler, 2011). In a recent study, however, enhancement of spatial working memory did not yield improved visuomotor adaptation in young adults (Anguera *et al.*, 2012). Thus, examining the relationship among different types of memory and motor tasks with different training-related parameters is important in future research.

Mental Imagery

Mental imagery is also known as visualization, mental practice or mental rehearsal. Highly skilled athletes tend to use mental imagery more often than novices for movement planning and motivational purposes before performance (Avinen-Barrow, Weigand, Hemmings, and Walley, 2008). Mental imagery is thought to prime one's movement initiation (Ramsey, Cumming, Eastough, and Edwards, 2010). More successful motor performance is often reported when imagery is used. Among tennis players, imagery helps increase serve accuracy and consistency (Guillot, Genevois, Desliens, Saieb, and Rogowski, 2012), and swimmers tend to swim faster after imagery training (Post, Muncie, and Simpson, 2012). Surgeons who mentally practice a virtual reality laparoscopic surgery task outperform those who do not (Arora *et al.*, 2011). Surprisingly, mental imagery also improves joint flexibility in both active and passive stretching, and the improvement is not associated with one's ability to imagine (Guillot, Tolleron, and Collet, 2010). Direction of imagery (imagining positive or negative consequences) is important, because it is related to performance motivation and self-efficacy (Short *et al.*, 2002). Negative imagery often debilitates motor performance, whereas positive imagery can be facilitative or neutral (Beilock, Afremow, Rabe, and Carr, 2001; Nordin and Cumming, 2005; Short *et al.*, 2002; Taylor and Shaw, 2002). Compatibility of sport and imagery type is also important. Different types of imagery recruit distinct neural networks for their processing, in which visual and kinesthetic imageries are mainly sub-served by the visual and motor-associated networks respectively (Guillot *et al.*, 2009). In sports where the form of movement is ultimately important (*e.g.* gymnastics, figure skating, diving), visual imagery is usually more effective than kinesthetic imagery to improve performance (Hardy and Callow, 1999). This finding implicates the significance of visual processing in sports where a premium is placed on the movement form or technique.

In motor skill learning, imagery can sometimes be used as a substitute or supplement to traditional physical practice for athletes of varying levels of expertise (Lacourse, Orr, Cramer, and Cohen, 2005). The value of mental imagery is greater for less experienced performers during early stages of learning (Bohan, Pharmer, and Stokes, 1999). In learning an object grasping and insertion task, the group with a combination of physical practice and mental imagery, and the control group that only practiced physically, showed improvement in movement times after practice. However, practice with more time allotted to mental rehearsal (50% and 70% of practice trials) resulted in better retention performance than physical practice alone (Allami, Paulignan, Brovelli, and Boussaoud, 2008). In a multiple-target arm-pointing task, both groups with physical practice and mental imagery showed similar retention performance, but the physical group learns faster (Gentili, Han, Schweighofer, and Papaxanthis, 2010). Motor imagery can also benefit the acquisition of spatio-temporal patterns of motor trajectories (Yágüez *et al.*, 1998). The learning advantages of employing motor training paradigms with a combination of mental imagery and physical practice have been noted in the sport of golf and table tennis (Bouziyne and Molinaro, 2005; Caliari, 2008). Compared with physical practice alone, a combination of physical practice and mental rehearsal is especially beneficial to the transfer of skill in different settings (Taktek, Zinsser, and St-John, 2008). Compatibility of sport and imagery type is crucial. Visual imagery is especially useful for acquiring tasks emphasizing the form of movement while kinesthetic imagery can more greatly benefit the timing control and the fine inter-limb coordination (Fery, 2003).

Imagery training has resulted in improved performance for certain types of athletes. For example, Calmels, Berthoumieux, and d'Arripe-Longueville (2004) find improved performance for sport-specific selective attention in softball players. High jumpers who have participated in imagery training improve their jumping height (Olsson, Jonsson, and Nyberg, 2008a). Examining training-related changes in the neural system during finger sequence learning, physical and imagery training would utilize different mechanisms to improve performance. Whereas physical training mainly involves the supplementary motor area and cerebellum, the visual cortex is involved in visual imagery training (Nyberg, Eriksson, Larson, and Marklund, 2006). A combination of physical and mental practice recruits both motor and visual systems. The benefits of training-related transfer are presumed to be a function of the connection between the cognitive and motor systems of the cerebellum

(Olsson, Jonson, and Nyberg, 2008b). In general, imagery training is more effective when administered individually, rather than to a group of people at the same time (Schuster *et al.*, 2011).

Cognitive Training and Motor Improvement

Training efficacy depends on both the effectiveness of the training program and cognitive plasticity (the trainees' capacity to acquire cognitive skills) (Baltes and Lindenberger, 1988; Calero and Navarro, 2007; Mercado, 2008). Those with greater cognitive plasticity are able to learn cognitive skills more readily. The effectiveness of one's cognitive plasticity is thought to be related to the experience-based neural plasticity (the structural and functional alternations of the human brain in response to the environmental stimulations), especially when presented with novel and demanding situations (Greenwood and Parasuraman, 2010). That is, when there is a prolonged difference between task demands and available cognitive resources, adaptive changes and cognitive plasticity occur in our cognitive system to narrow down the difference (Lövdén, Bäckman, Lindenberger, Schaefer, and Schmiedek, 2010). As demonstrated in the reported neuroimaging studies, prolonged external cognitive stimulations can impact our nervous system. For instance, in a longitudinal study, sustained navigation training helps preserve hippocampal integrity in both young and older adults (Lövdén *et al.*, 2012). In another study, subcortical structures is found to mediate the transfer of cognitive gains to untrained tasks after working memory training (Dahlin, Bäckman, Neely, and Nyberg, 2009).

Though people encounter cognitive and motor declines during aging or in later adulthood (Yan, 2000; Yan *et al.*, 2000), cognitive plasticity is relatively well preserved. Elderly people learn cognitive skills as well as their younger counterparts (Bherer *et al.*, 2006). However, when compared to children, older adults benefit less from mnemonic practice and attain lower final outcome scores than children, suggesting that children have greater memory plasticity than the elderly (Brehmer, Li, Müller, von Oertzen, and Lindenberger, 2007). As reported in previous studies, both young and older adults can benefit from various dual-task, memory, and navigation training, and sometimes show maintenance and positive transfer to untrained tasks (Bherer *et al.*, 2005; Bherer *et al.*, 2008; Cavallini, Pagnin, and Vecchi, 2003; Borella, Cerretti, Riboldi, and De Beni, 2010; McDougall and House, 2012). Although maintenance and positive transfer are sometimes reported, some studies show that with certain training paradigms, improved cognitive abilities are better preserved in young adults, and transfer is sometimes lacking in older adults (Dahlin, Nyberg, Bäckman , and Neely, 2008; Li *et al.*, 2008).

The spared cognitive plasticity is even evident among very old individuals (mean age of 80-years) (Buschkuel *et al.*, 2008). These individuals can maintain some of the training-induced cognitive enhancements for months (Yang and Krampe, 2009; Yang, Krampe, and Baltes, 2006). Older adults with lower working memory capacity seem to receive more benefits from working memory training than those of high capacity (Zinke, Zeintl, Eschen, Herzog, and Kliegel, 2012). However, existing training studies have tended to use small samples and different training paradigms. Thus, interpreting results is tenuous or difficult (Noack, Lövdén, Schmiedek, and Lindenberger, 2010; Zehnder, Martin, Altgassen, and Clare,

2009). Future research would benefit from larger samples. Researchers are also encouraged to devise more detailed theoretical frameworks regarding cognitive plasticity and enhancement. With more elaborate frameworks, more testable hypotheses regarding the training parameters and mechanisms can be posited. Results can be discussed within a common framework, so that meaningful discussions can be facilitated while our understanding of human plasticity is enhanced. One recently proposed theoretical framework (Lövdén *et al.*, 2010) explains the mechanisms by which cognitive training and plasticity are related, and the importance of adaptive training is implicated. This can direct future research and provide a common ground for discussion.

Cognitive training can improve a range of cognitive abilities in people of different ages, but how is cognitive training related to motor enhancement? Cognitive training-induced gains are often moderated by the cognitive plasticity of trainees. Those with greater cognitive plasticity have received more benefits from cognitive training. Because motor tasks have both cognitive and motor demands, cognitive training may lead to improvement in task performance via two separate routes as illustrated in Figure 1.

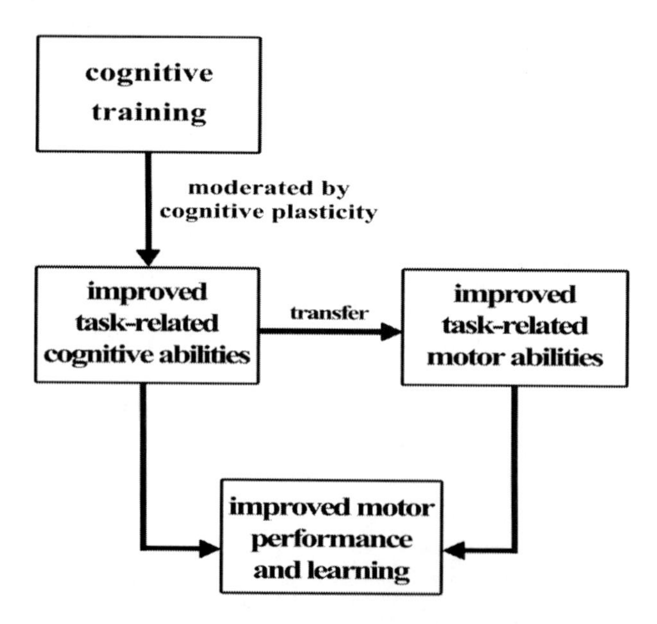

Figure 1. Illustration of how cognitive training may potentially benefit both cognitive and motor abilities, which in turn result in improved performance and learning of motor skills.

First, cognitive training can enhance the task-related cognitive abilities directly. Secondly, as described earlier in this chapter, there is dependence of motor performance and motor learning on cognition. Therefore, it is plausible that the training-induced cognitive gains can transfer to related motor abilities to a certain extent. Enhancement of motor performance and motor learning through cognitive training is provocative and plausible. Supporting evidence from laboratory research is accumulating, and future research is warranted to test its efficacy and applicability in real world settings.

Computerized Cognitive Training

Over the past few decades, there has been a dramatic increase in the processing capability of computers, and computers have become an indispensable part of daily living. Thus, many more computerized cognitive training programs are now available. This makes investigation of the training effects of computerized cognitive training a popular research area. In this section, we will learn about the efficacy of computerized cognitive training and examine its applicability to different groups of people.

Like traditional cognitive training, computerized training can augment cognitive performance. Computerized working memory training has been extensively researched and has generally been found to improve performance on working memory tasks. The training-related gains have even been shown to transfer to similar untrained tasks in both young and older adults (Brehmer, Westerberg, and Bäckman, 2012). The benefits of computerized cognitive training are not restricted to laboratory tasks. Recently, computerized cognitive training is shown to improve driving performance in a driving simulator following dual-task training for older adult participants (Cassavaugh and Kramer, 2009). This provides tentative and encouraging evidence that computerized cognitive training may enhance real world task performance.

Furthermore, clinical populations, such as patients with brain injuries, individuals with attention-deficit hyperactivity disorder (ADHD), and multiple sclerosis (MS), can also benefit from computer-based training programs (Holmes *et al.*, 2010; Klingberg, Forssberg, and Westerberg, 2002; Lundqvist, Grundström, Samuelson, and Ronnberg, 2010; Vogt *et al.*, 2009; Westerberg et al., 2007). Specifically, compared to traditional training, computerized training provides a more convenient way to enhance cognitive function. Computerized training can be administered at any place and time convenient to the trainee. This can be critical to trainees' compliance and training attendance rates. Moreover, trainers can track trainees' learning status and performance instantly. As a result of having constant feedback regarding the trainees' status and performance, the trainer has the advantage of offering prompt, or even instantaneous, feedback to participants or making adjustments to the training parameters as needed. Computerized cognitive training can be a viable, efficient, convenient and cost effective tool in augmenting one's cognitive and motor capabilities.

Research on computerized cognitive training programs is ongoing. Encouraging benefits from these kinds of cognitive training programs have been demonstrated repeatedly in various studies. Researchers are conceiving of new and different potential uses of computerized training for reasons other than cognitive training. Considerably more research on the application of computerized training on motor enhancement is likely in the future. However, a number of obstacles must be overcome to assure valid and reliable outcomes. One of the most important is standardizing training parameters, though no consensus has yet been established regarding parameters of an effective computer-based training program. Nevertheless, Klingberg (2010) recently proposes some guidelines. One of the most significant is assuring the adaptability of a training program (*i.e.* varying the difficulty as a function of the trainee's capability and performance). This is important as adaptable training programs can yield greater improvements than those that are non-adaptable (Brehmer, Westerberg, and Bäckman, 2012). Finally, the length of the training period is critical, but still unresolved. More research is needed to explore the optimal length of effective training programs.

Implications and Future Research

Potential applications of cognitive training may include improving motor performance of athletes, rehabilitation of patients with cognitive or motor deficits, and reversing age-related motor or cognitive declines. Athlete's performance is thought to improve when appropriate sport-related cognitive trainings are offered. In addition, some patients (*e.g.*, stroke, Alzheimer's disease) may have medical conditions that hinder their motor performance and learning. With cognitive training, these patients may learn, or relearn, motor skills that are critical for their daily function. Elderly people need the training to compensate for age-related cognitive or motor declines. Such training could be critical in fall and injury prevention and generally important for improving overall quality of life.

Like cognitive enhancement programs, understanding and applying cognitive training for motor enhancement is still not fully understood for several reasons. First, the specific cognitive abilities to be enhanced need to be identified so the cognitive training can specifically target the corresponding motor abilities. This requires more research to determine the links between a motor ability and the corresponding cognitive components. Secondly, parameters of cognitive training (*e.g.* training time, frequency, difficulty) require further investigation to establish the critical elements of the program. Thirdly, the ecological validity of cognitive training administered outside the laboratory remains questionable. Future research should address these issues to assure that cognitive training is sufficiently viable, practical, and available to the general public. Lastly, the issue of sport- or task-specificity needs additional attention. The cognitive abilities of expert and novice motor performers can usually be discriminated by sport- or task-specific measures; thus, for a training program to be effective, it has to be sport- or task-specific *per se* to improve the related cognitive and motor capabilities.

Conclusion

The idea that cognitive training can enhance motor performance and learning stems from the strong relationship between motor skills and human cognitive plasticity. Some promising benefits have been seen in research using cognitive training to enhance motor skill. Cognitive training provides an alternative, but highly plausible, way to improve motor control, learning, and development as a supplement to more direct physical practice and training. Using cognitive training (computerized and non-computerized) to improve motor skill is an increasingly popular notion that is still not fully understood. Further research is warranted to finalize conclusions regarding the effects of using cognitive training to augment selected cognitive abilities and, subsequently, the associated functional motor skills.

Acknowledgments

This work was supported in part by the Neuro-Academics Ltd. Hong Kong. There are no conflicts of interest.

References

Abbud, G.A.C., Li, K.Z.H., and DeMont, R.G. (2009). Attentional requirements of walking according to the gait phase and onset on auditory stimuli. *Gait and Posture, 30*(2), 227-232.

Abernethy, B., Baker, J., and Côté, J. (2005). Transfer of pattern recall skills may contribute to the development of sport expertise. *Applied Cognitive Psychology, 19*(6), 705-718.

Abernethy, B., Neal, R.J., and Koning, P. (1994). Visual-perceptual and cognitive differences between expert, intermediate, and novice snooker players. *Applied Cognitive Psychology, 8*(3), 185-211.

Abernethy, B., Schorer, J., Jackson, R.C., and Hagemann, N. (2012). Perceptual training methods compared: The relative efficacy of different approaches to enhancing sport-specific anticipation. *Journal of Experimental Psychology: Applied, 18*(2), 143-153.

Allami, N., Paulignan, Y., Brovelli, A., and Boussaoud, D. (2008). Visuo-motor learning with combination of different rates of motor imagery and physical practice. *Experimental Brain Research, 184*(1), 105-113.

Anguera, J.A., Bernard, J.A., Jaeggi, S.M., Buschkuehl, M., Benson, B.L., Jennett, S., Humfleet, J., Reuter-Lorenz, P.A., Jonides, J., and Seidler, R.D. (2012). The effects of working memory resource depletion and training on sensorimotor adaptation. *Behavioural Brain Research, 228*(1), 107-115.

Anguera, J.A., Reuter-Lorenz, P.A., Willingham, D.T., and Seidler, R.D. (2010). Contributions of spatial working memory to visuomotor learning. *Journal of Cognitive Neuroscience, 22*(9), 1917-1930.

Arora, S., Aggarwal, R., Sirimanna, P., Moran, A., Grantcharov, T., Kneebone, R., Sevdalis, N., and Darzi, A. (2011). Mental practice enhances surgical technical skills: A randomized controlled study. *Annals of Surgery, 253*(2), 265-270.

Avinen-Barrow, M., Weigand, D.A., Hemmings, B., and Walley, M. (2008). The use of imagery across competitive levels and time of season: a cross-sectional study among synchronized skaters in Finland. *European Journal of Sport Science, 8*(3), 135-142.

Baltes, P. B., and Lindenberger, U. (1988). On the range of cognitive plasticity in old age as a function of experience: 15 years of intervention research. *Behavior Therapy, 19*(3), 283–300

Beilock, S.L., Afremow, J.A., Rabe, A.L., and Carr, T.H. (2001). "Don't miss!" The debilitating effects of suppressive imagery on golf putting performance. *Journal of Sport and Exercise Psychology, 23*(3), 200-221.

Bherer, L., Kramer, A.F., Peterson, M.S., Colcombe, S., Erickson, K., and Becic, E. (2005). Training effects on dual-task performance: Are there age-related differences in plasticity of attentional control? *Psychology and Aging, 20*(4), 695-709.

Bherer, L., Kramer, A.F., Peterson, M.S., Colcombe, S., Erickson, K., and Becic, E. (2006). Testing the limits of cognitive plasticity in older adults: Application to attentional control. *Acta Psychologica, 123*(3), 261-278.

Bherer, L., Kramer, A.F., Peterson, M.S., Colcombe, S., Erickson, K., and Becic, E. (2008). Transfer effects in task-set cost and dual-task cost after dual-task training in older and younger adults: Further evidence for cognitive plasticity in attentional control in late adulthood. *Experimental Aging Research, 34*(3), 188-219.

Bohan, M., Pharmer, J.A., and Stokes, A.F. (1999). When does imagery practice enhance performance on a motor task? *Perceptual and Motor Skills, 88*(2), 651-658.

Borella, E., Cerretti, B., Riboldi, F., and De Beni, R. (2010). Working memory training in older adults: Evidence of transfer and maintenance effects. *Psychology and Aging, 25*(4), 767-778.

Bouziyne, M., and Molinaro, C. (2005). Mental imagery combined with physical practice of approach shots for golf beginners. *Perceptual and Motor Skills, 101*(1), 203-211.

Brehmer, Y., Li, S.C., Müller, V., von Oertzen, T., and Lindenberger, U. (2007). Memory plasticity across the life span: Uncovering children's latent potential. *Developmental Psychology, 43*(2), 465-478.

Brehmer, Y., Westerberg, H., and Bäckman, L. (2012). Working-memory training in younger and older adults: Training gains, transfer, and maintenance. *Frontiers in Human Neuroscience, 6*: 63.

Buschkuehl, M., Jaeggi, S.M., Hutchison, S., Perrig-Chiello, P., Dapp, C., Muller, M., Breil, F., Hoppeler, H., and Perrig, W.J. (2008). Impact of working memory training on memory performance in old-old adults. *Psychology and Aging, 23*(4), 743-753.

Calero, M.D., and Navarro, E. (2007). Cognitive plasticity as a modulating variable on the effects of memory training in elderly persons. *Archives of Clinical Neuropsychology, 22*(1), 63-72.

Caliari, P. (2008). Enhancing forehand acquisition in table tennis: The role of mental practice. *Journal of Applied Sport Psychology, 20*(1), 88-96.

Calmels, C., Berthoumieux, C., and d'Arripe-Longueville, F. (2004). Effects of an imagery training program on selective attention of national softball players. *Sport Psychologist, 18*(3), 272-296.

Cassavaugh, N.D., and Kramer, A.F. (2009). Transfer of computer-based training to simulated driving in older adults. *Applied Ergonomics, 40*(5), 943-952.

Cavallini, E., Pagnin, A., and Vecchi, T. (2003). Aging and everyday memory: The beneficial effects of memory training. *Archives of Gerontology and Geriatrics, 37*(3), 241-257.

Cereatti, L., Rasella, R., Manganelli, M., and Pesce, C. (2009). Visual attention in adolescents: Facilitating effects of sport expertise and acute physical exercise. *Psychology of Sport and Exercise, 10*(1), 136-145.

Chiviacowsky, S., Wulf, G., and Wally, R. (2010). An external focus of attention enhances balance learning in older adults. *Gait and Posture, 32*(4), 572-575.

Dahlin, E., Nyberg, L., Bäckman, L., and Neely, A.S. (2008). Plasticity of executive functioning in young and older adults: Immediate training gains, transfer, and long-term maintenance. *Psychology and Aging, 23*(4), 720-730.

Dahlin, E., Bäckman, L., Neely, A.S., and Nyberg, L. (2009). Training of the executive component of working memory: Subcortical areas mediate transfer effects. *Restorative Neurology and Neuroscience, 27*(5), 405-419.

Eccles, D.W., Walsh, S.E., and Ingledew, D.K. (2006). Visual attention in orienteers at different levels of experience. *Journal of Sports Sciences, 24*(1), 77-87.

Emanuel, M., Jarus, T., and Bart, O. (2008). Effect of focus of attention and age on motor acquisition, retention, and transfer: A randomized trial. *Physical Therapy, 88*(2), 251-260.

Fery, Y. (2003). Differentiating visual and kinesthetic imagery in mental practice. *Canadian Journal of Experimental Psychology, 57*(1), 1-10.

Furley, P., Memmert, D., and Heller, C. (2010). The dark side of visual awareness in sport: Inattentional blindness in a real-world basketball task. *Attention, Perception, and Psychophysics, 72*(5), 1327-1337.

Furley, P.A., and Memmert, D. (2010). Differences in spatial working memory as a function of team sports expertise: The corsi block-tapping task in sport psychological assessment. *Perceptual and Motor Skills, 110*(3), 801-808.

Furley, P.A., and Memmert, D. (2012). Working memory capacity as controlled attention in tactical decision making. *Journal of Sport and Exercise Psychology, 34*(3), 322-344.

Gentili, R., Han, C.E., Schweighofer, N., and Papaxanthis, C. (2010). Motor learning without doing: Trial-by-trial improvement in motor performance during mental training. *Journal of Neurophysiology, 104*(2), 774-783.

Greenwood, P.M., and Parasuraman, R. (2010). Neuronal and cognitive plasticity: A neurocognitive framework for ameliorating cognitive aging. *Frontiers in Aging Neuroscience, 2*: 150.

Guillot, A. Tolleron, C., and Collet, C. (2010). Does motor imagery enhance stretching and flexibility? *Journal of Sports Sciences, 28*(3), 291-298.

Guillot, A., Collet, C., Nyuyen, V.A., Malouin, F., Richards, C., and Doyon, J. (2009). Brain activity during visual versus kinesthetic imagery: An fMRI study. *Human Brain Mapping, 30*(7), 2157-2172.

Guillot, A., Genevois, C., Desliens, S., Saieb, S., and Rogowski, I. (2012). Motor imagery and 'placebo-racket effects' in tennis serve performance. *Psychology of Sport and Exercise, 13*(5), 533-540.

Hagemann, N., Strauss, B., and Canal-Bruland, R. (2006). Training perceptual skill by orienting visual attention. *Journal of Sport and Exercise Psychology, 28*(2), 143-158.

Hardy, L., and Callow, N. (1999). Efficacy of external and internal visual imagery perspectives for the enhancement of performance on tasks in which form is important. *Journal of Sport and Exercise Psychology, 21*(2), 95-112.

Holcombe, A.O., and Chen, W. (2012). Exhausting attentional tracking resources with a single fast-moving object. *Cognition, 123*(2), 218-228.

Holmes, J., Gathercole, S. E., Place, M., Dunning, D. L., Hilton, K. A., and Elliott, J. G. (2010). Working memory deficits can be overcome: Impacts of training and medication on working memory in children with ADHD. *Applied Cognitive Psychology, 24*(6), 827-836.

Holtzer, R., Wang, C., and Verghese, J. (2012). The relationship between attention and gait in aging: Facts and fallacies. *Motor Control, 16*(1), 64-80.

Hopwood, M.J., Mann, D.L., Farrow, D., and Nielsen, T. (2011). Does visual-perceptual training augment the fielding performance of skilled cricketers? *International Journal of Sports Science and Coaching, 6*(4), 523-535.

Kam, J.W.Y., Dao, E., Blinn, P., Krigolson, O.E., Boyd, L.A., and Handy, T.C. (2012). Mind wandering and motor control: Off-task thinking disrupts the online adjustment of behavior. *Frontiers in Human Neuroscience, 6*: 329.

Kioumourtzoglou, E., Derri, V., Tzetzis, G., and Theodorakis, Y. (1998). Cognitive, perceptual, and motor abilities in skilled basketball performance. *Perceptual and Motor Skills, 86*(3), 771-786.

Klingberg, T. (2010). Training and plasticity of working memory. *Trends in Cognitive Sciences, 14*(7), 317-324.

Klingberg, T., Forssberg, H., and Westerberg, H. (2002). Training of working memory in children with ADHD. *Journal of Clinical and Experimental Neuropsychology, 24*(6), 781-791.

Lacourse, M.G., Orr, E.L.R., Cramer, S.C., and Cohen, M.J. (2005). Brain activation during execution and motor imagery of novel and skilled sequential hand movements. *Neuroimage, 27*(3), 505-519.

Langan, J., and Seidler, R.D. (2011). Age differences in spatial working memory contribution to visuomotor adaptation and transfer. *Behavioural Brain Research, 225*(1), 160-168.

Lawrence, G.P., Gottwald, V.M., Hardy, J., and Khan, M.A. (2011). Internal and external focus of attention in a novice form sport. *Research Quarterly for Exercise and Sport, 82*(3), 431-441.

Li, S.C., Schmiedek, F., Huxhold, O., Rocke, C., Smith, J., and Lindenberger, U. (2008). Working memory plasticity in old age: Practice gain, transfer, and maintenance. *Psychology and Aging, 23*(4), 731-742.

Lohse, K.R., Sherwood, D.E., and Healy, A.F. (2010). How changing the focus of attention affects performance, kinematics, and electromyography in dart throwing. *Human Movement Science, 29*(4), 542-555.

Lohse, K.R., Sherwood, D.E., and Healy, A.F. (2011). Neuromuscular effects of shifting the focus of attention in a simple force production task. *Journal of Motor Behavior, 43*(2), 173-184.

Lövdén, M., Bäckman, L., Lindenberger, U., Schaefer, S., and Schmiedek, F. (2010). A theoretical framework for the study of adult cognitive plasticity. *Psychological Bulletin, 136*(4), 659-676.

Lövdén, M., Schaefer, S., Noack, H., Bodammer, N.C., Kuhn, S., Heinze, H., Duzel, E., Bäckman, L. and Lindenberger, U. (2012). Spatial navigation training protects the hippocampus against age-related changes during early and late adulthood. *Neurobiology of Aging, 33*(3), 620.e9-620.e22.

Lundqvist, A., Grundström, K., Samuelson, K., and Ronnberg, J. (2010). Computerized training of working memory in a group of patients suffering from acquired brain injury. *Brain Injury, 24*(10), 1173-1183.

Mann, D.T.Y., Williams, A.M., Ward, P., and Janelle, C.M. (2007). Perceptual-cognitive expertise in sport: A meta-analysis. *Journal of Sport and Exercise Psychology, 29*(4), 457-478.

Maxwell, J.P., Masters, R.S.W., and Eves, F.F. (2003). The role of working memory in motor learning and performance. *Consciousness and Cognition, 12*(3), 376-402.

McDougall, S., and House, B. (2012). Brain training in older adults: Evidence of transfer to memory span performance and pseudo-Matthew effects. *Aging, Neuropsychology, and Cognition, 19*(1-2), 195-221.

McNevin, N.H., Shea, C.H., and Wulf, G. (2003). Increasing the distance of an external focus of attention enhances learning. *Psychological Research, 67*(1), 22-29.

Memmert, D. (2007). Can creativity be improved by an attention-broadening training program? An exploratory study focusing on team sports. *Creativity Research Journal, 19*(2-3), 281-291.

Memmert, D. (2011). Creativity, expertise, and attention: Exploring their development and their relationships. *Journal of Sports Sciences, 29*(1), 93-102.

Memmert, D., Simons, D.J., and Grimme, T. (2009). The relationship between visual attention and expertise in sport. *Psychology of Sport and Exercise, 10*(1), 146-151.

Mercado, E. (2008). Neural and cognitive plasticity: From maps to minds. *Psychological Bulletin, 134*(1), 109-137.

Minvielle-Moncla, J., Audiffren, M., Macar, F., and Vallet, C. (2008). Overproduction timing errors in expert dancers. *Journal of Motor Behavior, 40*(4), 291–300.

Moore, L.J., Vine, S.J., Cooke, A., Ring, C., and Wilson, M.R. (2012). Quiet eye training expedites motor learning and aids performance under heightened anxiety: The roles of response programming and external attention. *Psychophysiology, 49*(7), 1005-1015.

Noack, H., Lövdén, M., Schmiedek, F., and Lindenberger, U. (2009). Cognitive plasticity in adulthood and old age: Gauging the generality of cognitive intervention effects. *Restorative Neurology and Neuroscience, 27*(5), 435-453.

Nordin, S.M., and Cumming, J. (2005). More than meets the eye: Investigating imagery type, direction, and outcome. *Sport Psychologist, 19*(1), 1-17.

Nyberg, L., Eriksson, J., Larsson, A., and Marklund, P. (2006). Learning by doing versus learning by thinking: An fMRI study of motor and mental training. *Neuropsychologia, 44*(5), 711-717.

Olsson, C.J., Jonsson, B., and Nyberg, L. (2008a). Internal imagery training in active high jumpers. *Scandinavian Journal of Psychology, 49*(2), 133-140.

Olsson, C.J., Jonsson, B., and Nyberg, L. (2008b). Learning by doing and learning by thinking: An fMRI study of combining motor and mental training. *Frontiers in Human Neuroscience, 2*:5.

Poolton, J.M., Maxwell, J.P., Masters, R.S.W., and Raab, M. (2006). Benefits of an external focus of attention: Common coding or conscious processing? *Journal of Sports Sciences, 24*(1), 89-99.

Poplu, G., Ripoll, H., Mavromatis, S., and Baratgin, J. (2008). How do expert soccer players encode visual information to make decisions in simulated game situations? *Research Quarterly for Exercise and Sport, 79*(3), 392-398.

Porter, J.M., Anton, P.M., and Wu, W.F.W. (2012). Increasing the distance of an external focus of attention enhances standing long jump performance. *Journal of Strength and Conditioning Research, 26*(9), 2389-2393.

Post, P., Muncie, S., and Simpson, D. (2012). The effect of imagery training on swimming performance: An applied investigation. *Journal of Applied Sport Psychology, 24*(3), 323-337.

Ramsey, R., Cumming, J., Eastough, D., and Edwards, M.G. (2010). Incongruent imagery interferes with action initiation. *Brain and Cognition, 74*(3), 249-254.

Ren, J., Wu, Y.D., Chan, J.S.Y., and Yan, J.H. (2013). Cognitive aging affects motor performance and learning. *Geriatrics and Gerontology International, 13*(1), 19-27.

Ridderikhoff, A., Peper, C.E., and Beek, P.J. (2008). Attentional loads associated with interlimb interactions underlying rhythmic bimanual coordination. *Cognition, 109*(3), 372-388.

Schack, T., and Mechsner, F. (2006). Representation of motor skills in human long-term memory. *Neuroscience Letters, 391*(3), 77-81.

Schucker, L., Hagemann, N., Strass, B., and Volker, K. (2009). The effect of attentional focus on running economy. *Journal of Sports Science, 27*(12), 1241-1248.

Schuster, C., Hilfiker, R., Amft, O., Scheidhauer, A., Andrews, B., Butler, J., Kischka, U., and Ettlin, T. (2011). Best practice for motor imagery: A systemic literature review on motor imagery training elements in five different disciplines. *BMC Medicine, 9*:75.

Seo, J., Kim, Y., Song, H., Lee, H.J., Lee, J., Jung, T., Lee, G., Kwon, E., Kim, J.G., and Chang, Y. (2012). Stronger activation and deactivation in archery experts for differential cognitive strategy in visuospatial working memory processing. *Behavioural Brain Research, 229*(1), 185-193.

Short, S.E., Bruggeman, J.M., Engel, S.G., Marback, T.L., Wang, L.J., Willadsen, A., and Short, M.W. (2002). The effect of imagery function and imagery direction on self-efficacy and performance on a golf-putting task. *Sport Psychologist*(1)*, 16*, 48-67.

Taktek, K., Zinsser, N., and St-John, B. (2008). Visual versus kinesthetic mental imagery: Efficacy for the retention and transfer of a closed motor skill in young children. *Canadian Journal of Experimental Psychology, 62*(3), 174-187.

Taylor, J.A., and Shaw, D.F. (2002). The effects of outcome imagery on golf-putting performance. *Journal of Sports Sciences, 20*(8), 607-613.

Taylor, J.A., and Thoroughman, K.A. (2007). Divided attention impairs human motor adaptation but not feedback control. *Journal of Neurophysiology, 98*(1), 317-326.

Totsika, V., and Wulf, G. (2003). The influence of external and internal foci of attention on transfer to novel situations and skills. *Research Quarterly for Exercise and Sport, 74*(2), 220-225.

Vine, S.J., and Wilson, M.R. (2011). The influence of quiet eye training and pressure on attention and visuo-motor control. *Acta Psychologica, 136*(3), 340-346.

Vogt, A., Kappos, L., Calabrese, P., Stöcklin, M., Gschwind, L., Opwis, K., and Penner, I. (2009). Working Memory training in patients with multiple sclerosis - Comparison of two different training schedules. *Restorative Neurology and Neuroscience, 27*(3), 225-235.

Voss, M.W., Kramer, A.F., Basak, C., Prakash, R.S., and Roberts, B. (2010). Are expert athletes 'expert' in the cognitive laboratory? A meta-analytic review of cognition and sport expertise. *Applied Cognitive Psychology, 24*(6), 812-826.

Ward, P., and Williams, A.M. (2003). Perceptual and cognitive skill development in soccer: The multidimensional nature of expert performance. *Journal of Sport and Exercise Psychology, 25*(1), 93-111.

Westerberg, H., Jacobaeus, H., Hirvikoski, T., Clevberger, P., Östensson, M. L., Bartfai, A., and Klingberg, T. (2007). Computerized working memory training after stroke - A pilot study. *Brain Injury, 21*(1), 21-29.

Williams, A.M., Davids, K., Burwitz, L., and Williams, J. (1993). Cognitive knowledge and soccer performance. *Perceptual and Motor Skills, 76*(2), 579-593.

Williams, A.M., Hodges, N.J., North, J.S., and Barton, G. (2006). Perceiving of play in dynamic sport tasks: Investigating the essential information underlying skilled performance. *Perception, 35*(3), 317-332.

Wulf, G., and Su, J. (2007). An external focus of attention enhances golf shot accuracy in beginners and experts. *Research Quarterly for Exercise and Sport, 78*(4), 384-389.

Wulf, G., Lauterbach, B., and Toole, T. (1999). The learning advantages of an external focus of attention in golf. *Research Quarterly for Exercise and Sport, 70*(2), 120-126.

Wulf, G., Shea, C.H., and Park, J. (2001). Attention and motor performance: Preferences for and advantages of an external focus. *Research Quarterly for Exercise and Sport, 72*(4), 335-344.

Wulf, G., Höß, M., and Prinz, W. (1998). Instructions for motor learning: Differential effects of internal versus external focus of attention. *Journal of Motor Behavior, 30*(2), 169-179.

Yágüez, L., Nagel, D., Hoffman, H., Canavan, A.G.M., Wist, E., and Homberg, V. (1998). A mental route to motor learning: Improving trajectorial kinematics through mental training. *Behavioural Brain Research, 90*(1), 95-106.

Yan, J.H. (2000). The effects of aging on linear and curvilinear arm movement control. *Experimental Aging Research, 26*, 393-407.

Yan, J.H., Thomas, J.R., and Payne, V.G. (2002). How children and seniors differ from adults in controlling rapid aiming arm movements. In J.E. Clark and J. Humphrey (Eds), *Motor development: Research and reviews* (vol. 2, pp 191-217). Reston, VA: NASPE.

Yan, J.H., Thomas, J.R. Stelmach, G.E., and Thomas, K.T. (2000). Developmental features of rapid aiming arm movements across the lifespan. *Journal of Motor Behavior, 32*, 121-140.

Yang, L., and Krampe, R.T. (2009). Long-term maintenance of retest learning in young, old, and oldest old adults. *Journal of Gerontology: Psychological Sciences, 64B*(5), 608-611.

Yang, L., Krampe, R.T., and Baltes, P.B. (2006). Basic forms of cognitive plasticity extended into the oldest-old: Retest learning, age, and cognitive functioning. *Psychology and Aging, 21*(2), 372-378.

Zehnder, F., Martin, M., Altgassen, M., and Clare, L. (2009). Memory training effects in old age as markers of plasticity: A meta-analysis. *Restorative Neurology and Neuroscience, 27*(5), 507-520.

Zinke, K., Zeintl, M., Eschen, A., Herzog, C., and Kliegel, M. (2012). Potentials and limits of plasticity induced by working memory training in old-old age. *Gerontology, 58*(1), 79-87.

In: Motor Behavior and Control: New Research
Editors: Marco Leitner and Manuel Fuchs
ISBN: 978-1-62808-142-8

Chapter 4

Laterality, Load, and Motor Imagery

*Andrew B. Slifkin**
Cleveland State University, Cleveland, OH, US

Abstract

Actual and imagined movement durations (MDs) are equivalent when inertial loads are low relative to the maximum inertial loading capacity of (dominant) index finger movement, but as load increases imagined MDs lengthen at a faster rate than actual MDs (Slifkin, 2008). The load-dependent lengthening of imagined over actual MD may arise from increased uncertainty in predicting motor output: Uncertainty in predicting movement outcomes has been correlated with increases in MD. Here, the non-dominant and dominant index fingers participated in actual and imagined action under different loads. Because the non-dominant finger may have less experience moving heavier loads—with increased uncertainty in controlling those loads—it was hypothesized that the actual-imagined MD gap should increase at a faster rate for the non-dominant finger. The results for both fingers replicated prior research (Slifkin, 2008). However, there were no laterality effects. Movement in the current task was limited to a single biomechanical degree of freedom. A possible explanation for the absence of a laterality effect is that the non-dominant and dominant controllers are equally effective in controlling systems with few elements.

Evidence suggests that common cognitive and neural processes operate during actual and imagined action (see Jeannerod, 2006, for a review). Some of that evidence comes from studies showing that actual and imagined movement durations (MDs) are equivalent under conditions of effector loading (Gentili, Cahouet, Ballay, & Papaxanthis, 2004; Papaxanthis, Schieppati, Gentili, & Pozzo, 2002). However, other studies have shown that effector loading results in imagined MDs exceeding actual MDs (Cerritelli, Maruff, Wilson, & Currie, 2000;

* Corresponding author: Department of Psychology, Cleveland State University, 2121 Euclid Avenue, Cleveland, OH 44115, USA, E-mail: a.slifkin@csuohio.edu.

Decety, Jeannerod, & Prablanc, 1989). How can we account for the differences in results among those studies?

A review of the literature on load and motor imagery (Slifkin, 2008) indicated that when studies used loads that appeared low relative to the estimated maximum capacity of the effector system, the actual-imagined MD equivalence was preserved (Gentili et al., 2004; Papaxanthis et al., 2002), but when studies used loads that appeared high relative to the estimated maximum loading capacity of the effector system, imagined MDs exceeded actual MDs (Cerritelli et al., 2000; Decety et al., 1989). For example, in one study, the actual-imagined MD equivalence was preserved for shoulder movement loaded at 1.5 kg (Papaxanthis et al., 2002), but in another study, large actual-imagined MD differences were found for wrist movements loaded at 2 kg (Cerritelli et al., 2000). While the absolute values of the loads in both studies were close, wrist movement with a 2-kg load should be closer to the maximum capacity of the wrist than shoulder movement with a 1.5-kg load would be to the maximum capacity of the shoulder. In other words, if the shoulder and wrist effector systems are required to move loads of similar absolute values and it is reasonable to assume that shoulder strength is in excess of wrist strength, then the wrist can be said to operate closer to its maximum capacity than the shoulder would operate relative to its maximum capacity. Support for the notion that the shoulder is stronger than the wrist comes from the finding that the maximum force production capacity of the wrist is much less than that of the shoulder (Stoll, Huber, Seifert, Michel, & Stucki, 2000).[1] Thus, the presence or absence of actual-imagined MD differences might depend on the proximity of the load to the maximum capacity of the engaged effector system.

To test the hypothesis that the presence and magnitude of actual-imagined MD differences should increase as load increases toward the maximum loading capacity of the engaged effector system, Slifkin (2008) introduced a procedure for assessing the maximum voluntary loading capacity [maximum voluntary load (MVL)] of single degree of freedom movement of the index finger. That procedure used a systematic, trial-by-trial adjustment of loads to home in on the MVL. Once a participant's MVL was identified, loads were scaled to different percentages of the MVL [0, 5, 10, 20, 40, & 80% MVL] and participants had to actually move or imagine moving their dominant (right) index finger in a cyclical aiming task (see Figure 1). The results showed that the actual and imagined MDs did not differ at low loads; however, with further increases in the load requirement both actual and imagined MDs increased, but imagined MDs increased at a faster rate. In other words, the results showed that the actual-imagined MD equivalence was preserved at the lower loads, but as load increased, actual-imagined MD differences increased. Those results supported the study's hypothesis that the presence and magnitude of actual-imagined MD differences should increase as load increases toward the maximum loading capacity of the effector system.

A possible explanation of the increases in *both* actual and imagined MDs under high-load conditions is that participants have less everyday experience in moving heavy loads, which would relate to less complete central representations for high-load action, and, when that

[1] Stoll et al. (2000, Table 3, p. 108) provides the maximum isometric force production values for a range of effectors and different actions within each effector. When averaged across values for males and females and across force production during flexion and extension, the maximum force production values were 257.51 N for the shoulder and 163.58 N for the wrist. Thus, the maximum strength of the shoulder is about 1.6 times that of the wrist. [Of the various shoulder and wrist actions examined by Stoll et al. (2000), the extension and flexion actions were most similar to the wrist movements examined by Cerritelli et al. (2000) and the shoulder movements examined by Papaxanthis et al. (2002).

information is less complete there is greater uncertainty in predicting if and how movement can reach its goals (Slifkin, 2008). Given prior evidence that uncertainty in predicting movement outcomes correlate with lengthened MDs (Garner, 1962; Schmidt, Zelaznik, Hawkins, Frank, & Quinn, 1979; Carson, Elliot, et al., 1993), the high-load lengthening of both actual and imagined MDs might reflect compensation for uncertainty in predicting action outcomes—according to a *speed-uncertainty trade-off* (Slifkin, 2008). When there is high uncertainty in predicting movement outcomes, planning movement to be of long duration—even longer than actually necessary—may reflect a safe, conservative strategy, i.e., it is better to be safe than sorry (Slifkin, 2008). In other words, achievement of movement outcome success should be maximized when the planned MDs overestimate rather than underestimate what is required.[2]

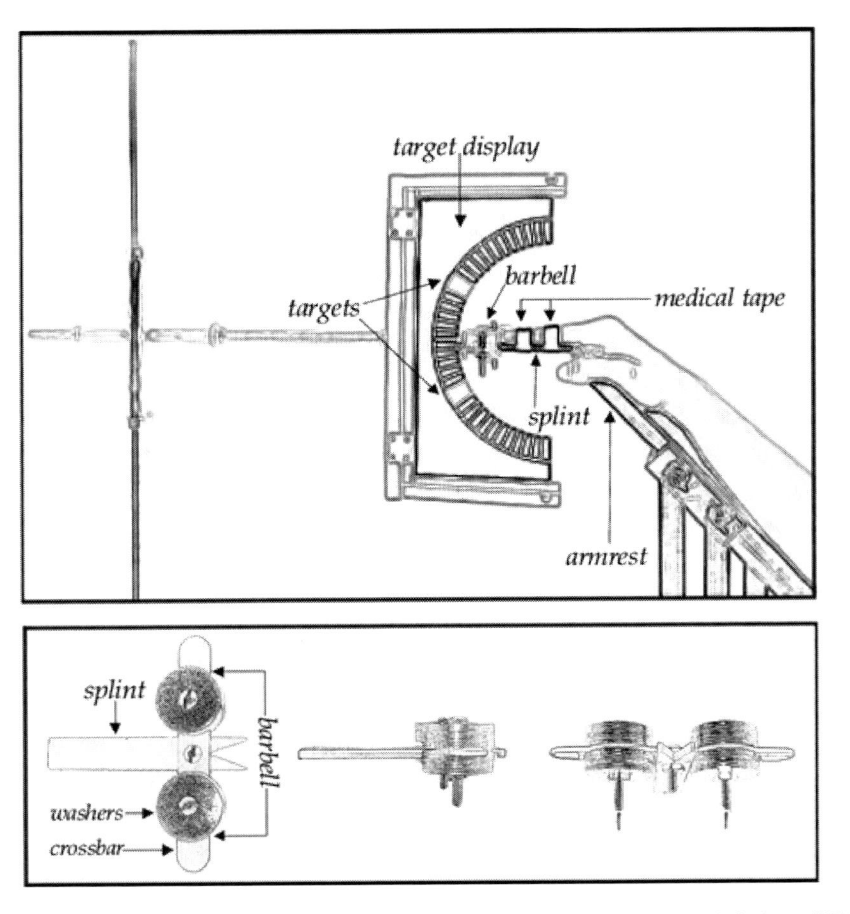

Figure 1. The target display, armrest, and inertial loads. Reprinted from "High loads induce differences between actual and imagined movement duration," by A. B. Slifkin, 2008, *Experimental Brain Research*, *185*, p. 300, Copyright 2008 by Springer-Verlag. Reprinted with permission.

[2] One factor that may contribute to age-related slowing of movement is the adoption of a safe, conservative strategy where MDs lengthen in order to maximize movement success, or, in other words, reduce uncertainty in predictions of movement outcomes (e.g., Rabbit, 1979; Welford, 1984). Increases in uncertainty of movement outcomes could be related to age-related motor-system degradation.

Thus, the load-induced increases in *both* actual and imagined MD might result from compensation for outcome uncertainty. However, how can the load-induced increases in the actual-imagined MD gap be accounted for in the results of Slifkin (2008) and other studies (Cerritelli et al., 2000; Decety et al., 1989)? There is good support for the hypothesis that the information processes associated with the planning and preparation of actual action are the *same* processes engaged during imagined action (Jeannerod, 1994). Further, according to the framework provided by internal models of action (e.g., Desmurget & Grafton, 2000), the instructions for internal simulations of action—i.e., imagined action—are based on copies of the initial feedforward commands destined to drive actual action (Wolpert, Ghahramani, & Jordan, 1995; Wolpert, & Ghahramani, 2000). Given that the signals for actual and imagined action are the same, actual and imagined MDs should be the same too. Indeed, equivalent actual and imagined MDs have frequently been observed in motor imagery research (e.g., Jeannerod, 2006).

Such matching of MDs might be expected when actors perform actions with highly predictable outcomes such as those with which they are highly familiar. In addition, highly familiar and predictable movements should largely be governed by open-loop, feedforward motor control processes *without* reliance on online sensorimotor feedback that can be used to adjust and amend actual movement trajectories as they unfold—viz., through closed-loop visual feedback processes (e.g. Woodworth, 1899; Meyer, Abrams, Kornblum, Wright, & Smith 1988). On the other hand, under unfamiliar high-load conditions where central representations for action are not as strong and there is greater uncertainty in predicting movement outcomes, more lengthy MDs may initially be planned. When those plans are used to simulate imagined action, the lengthy imagined MDs should be of equal duration to the *plans* for actual action. However, when those plans for action are actuated, the availability of online sensorimotor feedback might allow movement to be completed in a shorter period than specified by the original feedforward commands. Therefore, at high-load requirements, actual MDs would be of shorter duration than imagined MDs.

Like the study by Slifkin (2008), the current study examines changes in actual and imagined MDs under conditions of load (0 & 80% MVL). Here, however, performance was examined for both the non-dominant and dominant index fingers of right-hand dominant individuals.

There is evidence that the non-dominant hand has reduced capacity for the control of inertial loads: Carson and colleagues (Carson, Elliot, et al., 1993) examined actual handheld stylus movements under different load conditions and found [that MD increased with load (0, 1,120, and 2,200 g) and] that MD was longer for the non-dominant hand. Other studies have demonstrated performance advantages for the dominant hand of right-handed individuals (e.g., Bagesteiro & Sainburg, 2003; Peters & Durding, 1979; Provins & Magliaro, 1989; Roy, Kalbfleisch, & Elliott, 1994). In addition, it is known that there are larger volumes and densities of neural structures (axons, dendrites, synapses) within the hand areas of the contralateral sensory and motor cortices (Amunts et al., 1996; Jung et al., 2003; Sörös et al., 1999).

Taken together, such evidence implies that there should be enhanced capacity for the control of action with the dominant hand, which may relate to stronger central representations for dominant hand action.[3] Thus, in the current study, it is predicted that there will be greater uncertainty in the ability of the non-dominant index finger to control its output when moving loads, and, in particular, heavier loads. That should create longer plans for actual MDs specified by the initial feedforward commands for action, longer internal simulations, and, therefore, longer imagined MDs for the non-dominant finger.

Because there should be greater uncertainty for controlling heavy loads with the non-dominant effector, an expectation is that the load-induced increases in imagined MDs would be sharper for the non-dominant finger: Again, the overly lengthened initial feedforward commands for action are copied and used to produce an internal simulation of action—i.e., the imagined movement (Desmurget & Grafton, 2000); such over-lengthening of the non-dominant imagined MDs might reflect a safe, conservative strategy to increase the likelihood of success in the face of uncertainty (Slifkin, 2008). However, when actual movement is executed, the availability and engagement of sources of online feedback may enable a reduction of the actual MD below the level reflected in the imagined MD. For high-load movement with the non-dominant as opposed to the dominant finger, there should be more uncertainty to reduce as reflected in longer non-dominant imagined MDs. Thus, there should be larger feedback-based imagined-to-actual MD reductions for high-load non-dominant movement. In other words, it is predicted that the load-dependent gap between actual and imagined MDs should grow at a faster rate for the non-dominant index finger as compared with the dominant index finger. Testing that hypothesis was the main purpose of the current study.

Method

Participants

The participants in the experiment were selected from a larger group of individuals that completed the Edinburgh Handedness Inventory (EHI) (Oldfield, 1971) in undergraduate, university classrooms. The EHI provides a laterality quotient (LQ) that ranges from -100 to +100, and reflects, respectively, the extremes of left-hand and right-hand dominance. Those from the initial group with high, positive LQ scores were asked to participate in the study. Of that group, 11 experimentally naïve, healthy individuals agreed to participate. Ten participants were female. The mean LQ score for the group was 90.14 (*SD* = 10.71) and their mean age was 23.71 years (*SD* = 6.90). The local institutional review board approved the experiment, and each participant provided informed consent at the start of the experiment. Upon completion of the experiment, each participant received 20 USD.

[3] In addition, experience may contribute to enhanced capacity of the dominant hand, in general, and the control of high loads with the dominant hand, in particular. Participants in the current study were strongly right-hand dominant and their hand dominance was assessed using the Edinburgh Handedness Inventory (Oldfield, 1971). It includes questions about hand use in tasks where participants are asked about their interaction with a range of everyday, normally weighted objects. Had those objects been heavily weighted, then it seems reasonable to assume that participants would continue to report a preference for using their dominant hand. Thus, as compared with the non-dominant hand, the dominant hand should gain greater experience handling both normally weighted objects and unusually heavy objects. Consequently, there should be stronger central representations for heavily loaded dominant hand action.

Apparatus

Aside from testing the non-dominant index finger, the experimental set-up and procedure were essentially the same as those reported by Slifkin (2008). Additional detail is available in that report.

As shown in Figure 1, the apparatus consisted of a target display, an armrest, and weights used to load the index finger. In general, during each experimental trial, participants were required to move or to imagine moving their loaded finger between two targets in a cyclical aiming task. Each target was 10° in width and was placed along a semicircle so that the target midpoints were located at -30° and +30° from the midpoint of the semicircle.[4] The semicircle midpoint served as the 0° mark, and when participants aligned the end of a splint attached to the finger at 0°, the index finger was in a horizontal position. Movement of the index finger occurred in the sagittal plane, and the finger splint restricted movement to a single biomechanical degree of freedom at the metacarpophalangeal (MCP) joint. The armrest supported the forearm so that it was maintained at 45° with respect to the surface of the table on which the armrest was attached. The loads consisted of 32 *barbells*, and they ranged in weight from 2.5 to 550 g. The loads were used to determine a participant's MVL for the left and right index finger. Then, the barbell load closest to 80% of each participant's MVL value was selected as the 80% MVL load. Under all 0% MVL conditions, only the splint was attached to the finger.

Procedure

Assessement of the maximum voluntary load. Prior to the start of MVL assessment, detailed instructions were given about the movement task. At the beginning of each trial, participants positioned their index finger at the 0° mark on the target display, and they rested their untested hand in their lap and held a wireless mouse. Then, with the finger being tested, they moved the pointed end of the splint (the pointer) toward the lower target (flexion). Once the pointer was within its bounds, a smooth movement reversal was made towards the upper target (extension). The mouse button was pressed at the moment of the first reversal, and it activated a computer-program timer that recorded the MD for each trial, with 1-ms resolution. The first entry into the upper target region was considered the first target contact, and the sequence of up-and-down movements continued until the pointer crossed the boundary of the sixth target in the sequence. It was at that moment that participants pressed the mouse button again to stop the timer. The interval between the first and second button press was the MD.[5] It was emphasized that performance should occur at a comfortable pace (Decety et al., 1989;

[4] Thus, with the distance (D) between target centers set at 60° and target width (W) set at 10°, according to Fitts' index of difficulty (ID)—$\log_2(2D/W)$—the ID in the current experiment took on a value of 3.58 bits, which could be considered a moderate, if not moderately high, ID value (e.g., see Table 3 in Plamondon & Alimi, 1997).

[5] In the current experiment, button presses were made with the "non-performing" index finger to mark the "performing" index finger's movement initiation and then movement termination; the interval between those events was the MD for the performing finger. Such methodology matches that used previously in the study of actual and imagined movement of the left and right arm by Skoura, Personnier, Vinter, Pozzo, & Papaxanthis (2008, p. 1273). Further, those investigators conducted studies that compared their method to other methods for recording actual and imagined MDs: They found all methods to be equivalent.

Gentili et al., 2004; Papaxanthis et al., 2002), and that target-to-target movements and direction reversals should be performed smoothly. Successful trials were defined as those including six continuous target-to-target movements, whereas unsuccessful trials were defined as those where fewer than six target-to-target movements were completed, or those where movements were noticeably discontinuous, or those that contained prolonged pauses. Movement duration was the only dependent variable recorded in the current study.

The MVL estimation procedure used an iterative adjustment of the barbell weight, and an associated narrowing of the range of candidate barbells to home in on the MVL. The procedure always began with participants lifting the 16th barbell in the range of 32 barbells. If an MVL trial was successful, then there was an increase in barbell weight on the next trial, but if a trial was unsuccessful, then the weight was reduced. In addition, with each adjustment, half of the barbells were retained as MVL candidates, and the other half were eliminated from consideration. The trial-by-trial halving of the range of barbells continued until the MVL was located (for more detail see Slifkin, 2008).

Experimental instructions. Before the start of the experiment, participants were told that an actual movement trial would be performed according to the instructions for an MVL trial. Under both the actual and imagined 80% MVL performance conditions, barbell loads were attached to the splints, and under both 0% MVL performance conditions only the splint was attached to the finger. Unlike the actual movement conditions, throughout each imagined trial participants maintained their finger at the 0° starting position, and did not generate any movement. Before the start of the imagined trials, participants were asked to view the target display and their loaded finger for several moments, so that when they closed their eyes they would form a vivid, detailed image of the scene. Participants then imagined moving their finger from the 0° position to the lower target. When the pointer, in imagination, crossed the inner boundary of the lower target, they were to press the mouse button to start the timer, and at the moment the imagined movement crossed the inner boundary of the sixth target, they were to press the mouse button again to stop the timer.

Experimental Design and Data Analyses

The left and right fingers were tested during separate halves of the experimental session. The finger tested during the first half of the experiment was counterbalanced across participants. During each half of the session, the tested finger's MVL was estimated and then the experimental data were collected. For each finger, participants completed four series of four trials where each series included an actual and an imagined trial at each of two load conditions. The order of trials within each series was completely randomized.

The initial data analysis provided a test of differences in the maximum loading capacity of the non-dominant and dominant index fingers. That analysis used a two-tailed dependent-sample t test to compare the MVL values for the two fingers. (The statistical outcome of the MVL comparison would be the same as that of a t test used to compare the two 80% MVL load requirements.) Then, the MD data from the main part of the experiment was analysed with a three-way index finger (left, right) by performance condition (actual, imagined) by load (0, 80% MVL) repeated-measures ANOVA. The data entered into the ANOVA for each participant were averages across the four series at each of the eight unique conditions within the experimental design. According to the results of Slifkin (2008), there should be significant main effects for performance condition (imagined > actual MD) and load (0 < 80% MVL MD), and an interaction of those factors showing an increasing actual-imagined MD gap from 0 to 80% MVL. The main prediction of the current study of larger load-dependent increases in the actual-imagined MD gap for the non-dominant finger would be revealed by a significant three-way interaction. In particular, the actual-imagined MD gap should increase at a faster rate as a function of load for the non-dominant finger as compared to the dominant finger. The prediction that the non-dominant finger should generally be slower than the dominant finger would result in a significant main effect for index finger.

Results

The group-mean MVL values for the non-dominant and dominant finger were 309.45 g (*SD* = 61.45) and 327.45 g (*SD* = 70.76 g), respectively. While the MVL value was less for the non-dominant finger, the difference was not reliable, $t(10) = -1.33, p > .05$. In essence, the maximum loading capacities for the non-dominant and dominant index fingers were equivalent. [Likewise, the group-mean 80% MVL values were equivalent for the non-dominant (*M* = 247.56 g, *SD* = 49.16 g) and dominant finger (*M* = 261.96 g, *SD* = 56.61 g), $t(10) = -1.33, p < .05$.][6]

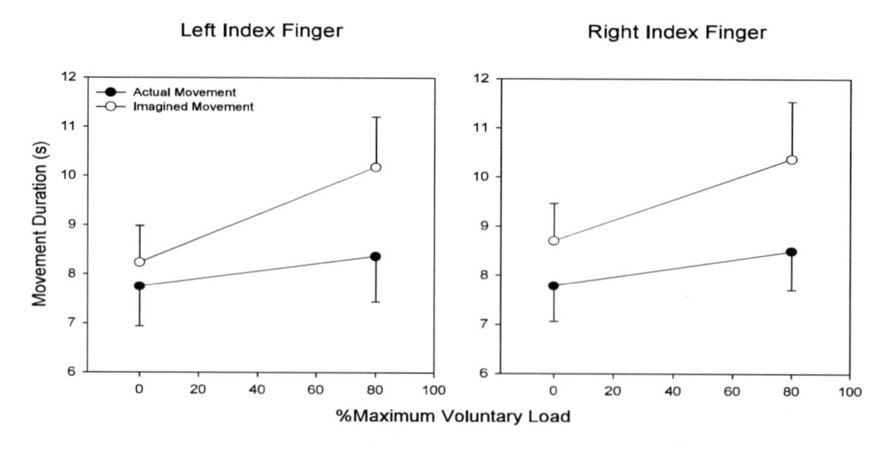

Figure 2. Changes in the average actual and imagined movement duration (MD) as a function of percent maximum voluntary load (%MVL) for the left (non-dominant) and right (dominant) index fingers. Each data point reflects the group-mean MD under each unique condition, and error bars represent one standard error of the mean (*SEM*). Here, the *SEM* has been plotted above each imagined MD data point (+1 *SEM*) and below each actual MD data point (-1 *SEM*).

[6] The absence of a reliable laterality effect for the MVL differs from the majority of studies on maximum force production. Largely, those studies seem to show that the dominant hand and fingers are stronger than the non-dominant hand and fingers (e.g., Henningsen, Ende-Henningsen, & Gordon, 1995; Incel, Ceceli, Durukan, Erdem, & Yorgancioglu, 2002; Kamarul, Ahmad, & Loh, 2006). However, as discussed further in the Discussion, there appears to be a tendency for the magnitude and consistency of the laterality effect to decline as the number of digits or degrees of freedom involved in force production declines. In the current study, movement was restricted to a single biomechanical degree of freedom of the index finger—the MCP joint. In addition, it should be noted that estimates of the maximum capacity of an effector system have typically been obtained using isometric force production tasks. In contrast, the current MVL procedure required movement of inertial loads in a cyclical aiming task.

Figure 2 shows changes in the group-mean MDs as a function of %MVL under the actual and imagined movement conditions for the non-dominant (left panel) and dominant (right panel) index fingers. An initial observation is the close similarity in the pattern of results for both fingers: For both fingers, and within both the actual and imagined conditions, MD increased from 0 to 80% MVL. Those load-dependent increases were sharper for the imagined conditions, which gave rise to larger actual-imagined MD gaps at 80% as compared with 0% MVL.

The observations from Figure 2 were supported by results from the three-way index finger (2) by performance condition (2) by load (2) ANOVA: First, there was a reliable main effect for performance condition showing that imagined MDs ($M = 9.37$ s, $SD = 2.83$ s) were longer than actual MDs ($M = 8.10$ s, $SD = 2.50$ s), $F(1, 10) = 13.23$, $p < .01$. Second, a significant main effect for load indicated that MDs reliably lengthened from 0% MVL ($M = 8.12$ s, $SD = 2.36$ s) to 80% MVL ($M = 9.35$ s, $SD = 2.91$ s), $F(1, 10) = 18.08$, $p < .01$. Third, the observation that imagined MDs lengthened at a faster rate than the actual MDs as a function of load—and therefore the magnitude of the actual-imagined MD gap increased with increases in %MVL—was supported by a significant performance condition by load interaction, $F(1, 10) = 13.61$, $p < 0.01$. That interaction replicates the main findings of Slifkin (2008).[7] Fourth, the overall group-mean MD for the left ($M = 8.63$ s, $SD = 2.84$) and right ($M = 8.84$ s, $SD = 2.75$ s) finger was quite similar, and the main effect for finger was absent, $F(1, 10) = .12$, $p = .74$. In addition, there were no reliable interactions with finger (ps > .05).

Discussion

In the current study, it was predicted that the actual-imagined MD difference should increase as a function of load at a faster rate for the non-dominant as compared with the dominant index finger. A secondary prediction might be that the non-dominant index finger would generally be slowed when compared with the dominant finger. While the current results replicated and extended the main findings of Slifkin (2008), there were no laterality effects. The remainder of the Discussion will examine factors that may have influenced the absence of the predicted laterality effects.

Task Difficulty and Laterality Effects

Other studies have identified conditions under which laterality effects are absent (Bryden, Roy, Rohr, & Egilo, 2007; Flowers, 1975) and conditions where performance of the non-dominant effector is superior to the dominant effector (Bagesteiro and Sainburg, 2002; Bagesteiro and Sainburg, 2003; Sainburg, 2005; Freitas, Krishnan, & Jaric, 2007;

[7] The Scheffé post-hoc test was used to compare means from the performance condition by load interaction of the three-way ANOVA. In particular, it was of interest to examine the reliability of the differences in the actual and imagined MDs within each level of load. According to the results of Slifkin (2008), reliable actual-imagined MD differences should be absent at 0% MVL and should appear at 80% MVL. In the current data, at 0% MVL there was some elevation of the group-mean imagined MD over the group-mean actual MD, but that difference was not reliable according to the Scheffé post-hoc test ($p > .05$). In contrast, at 80% MVL there was a highly reliable elevation of the group-mean imagined MD over the group-mean actual MD ($p < .001$).

Schabowsky, Hidler, & Lum, 2007). However, when MD was a main dependent variable in studies on targeted aiming—as it was here—the results have largely shown that MDs for the non-dominant hand were longer than those of the dominant hand (Flowers, 1975; Roy & Elliott, 1986; Roy & Elliott, 1989; Carson, Goodman, et al., 1993; Carson, Elliott, et al., 1993; Roy, Kalbfleisch, & Elliott, 1994; Hoffman, 1997).

One factor that may have influenced the appearance of the laterality effects in those studies was the imposed movement difficulty, viz., as defined by Fitts' index of difficulty (ID) (see Footnote 4). In those studies, the movement distance requirements were high (20 to 40 cm) and the target was either a point, formed by the intersection of horizontal and vertical lines (Roy & Elliott, 1986; Roy & Elliott, 1989; Carson, Goodman, et al., 1993; Carson, Elliott, et al., 1993), or the target was defined but narrow (Roy et al., 1994: 1 cm target width). Because of the narrow targets and long movement distances, the IDs would be quite high. Other studies used a cyclical aiming task where ID was explicitly varied: One study showed that MD was longer for the non-dominant hand at all IDs (1 to 6 bits), with differences reaching statistical significance at 4 bits and higher (Flowers, 1975). A more recent cyclical aiming study showed that MDs for the non-dominant hand were statistically longer than those for the dominant hand at all IDs (2 to 6 bits) (Hoffman, 1997).

In the current experiment—which used a cyclical aiming task—the ID was set at 3.58 bits (see Footnote 4), which is quite close to the ID level where a reliable laterality effect was found in Flowers (1975) and it was within the range of IDs where reliable laterality effects were found in Hoffman (1997). Thus, with the ID level used in the current experiment, it was reasonable to expect laterality effects for the actual MDs, even under the 0% MVL (no load) condition. Even if the effect did not appear at 0% MVL, the expectation of an effect at 80% MVL would seem to increase. That is, the 80% MVL load should place a greater tax on the capacity of the non-dominant index finger, as compared with the tax placed on the dominant index finger.

Degrees of Freedom, Task Complexity, and Laterality Effects

Regardless of the appearance of a laterality effect for the actual MDs at either load level (0% or 80% MVL), it was reasonable to assume that there would be greater uncertainty in the control of heavier loads for the non-dominant finger and, consequently, that the imagined MDs for the non-dominant finger should exceed the imagined MDs for the dominant finger, especially under heavy loads. That notion was based on the assumption that the non-dominant finger would have less experience moving heavy loads and, therefore, would have weaker central representations for moving them (see Footnote 3). During actual movement, the expression of increased uncertainty for the non-dominant index finger could have been compensated for through the engagement of online sensorimotor feedback to guide the finger to its target (e.g., Woodworth, 1899; Meyer, Abrams, Kornblum, Wright, & Smith 1988); participants may have made quick use of such feedback, which could have allowed for the discovery that the goal could be achieved with a briefer MD than originally planned. In that case, the absence of a laterality effect for actual MDs might have been expected. However, online sensorimotor feedback is absent during motor imagery. In that case, more lengthy imagined MDs at 80% MVL might have been expressed for the non-dominant index finger as compared with the dominant index finger. Thus, if at 80% MVL the actual MD for the non-dominant and dominant hands were equivalent, but at 80% MVL the imagined MD for the non-dominant finger was longer than the imagined MD for the dominant finger, then the actual-imagined MD difference would be greater at 80% MVL for the non-dominant index finger. That prediction was not realized.

Could it be that the chances of finding performance superiority of the dominant effector increases with the need to recruit an increasing number of degrees of freedom to meet task demands? Futher, could it be that increases in task complexity or task demands are correlated with engagement of an increased number of biomechanical degrees of freedom? For example, handwriting may be the prime example of a task where there are strong laterality effects. It requires the production of complex geometric forms that may involve sequential activation of multiple biomechanical degrees of freedom across the fingers, hand, elbow and even shoulders. In contrast, in tasks like the production of maximum force output using a power grip—involving simultaneous activation of all muscles of the hand—the dominant hand advantage is considerably less than that of handwriting (Provins & Magliaro, 1989). The dominant hand advantage for the power grip appears to lessen further when fewer digits participate in force production, viz., when the thumb and index finger generate force via a pinch grip (Kunelius, Darzins, Cromie, & Oakman, 2007). In addition, in the studies on discrete targeted aiming that had consistently shown shorter MDs for the dominant hand, it seems likely that the large movement distance requirements (20 to 40 cm) would have engaged all joints of the upper extremity (Roy & Elliott, 1986; Roy & Elliott, 1989; Carson, Goodman, et al., 1993; Carson, Elliott, et al., 1993; Roy et al., 1994). For example, activation of the shoulder and elbow joints might be responsible for covering most of the distance to the target and engagement of the wrist and fingers might be needed for finer adjustments to the narrow targets used in those studies. In sum, it appears that when more biomechanical degrees of freedom need to be recruited and coordinated to meet task demands, the strength of laterality effects increase. Next, a more detailed example will be provided of how task complexity and engagement of biomechanical degrees of freedom might relate to laterality effects.

Recently, Bryden et al. (2007) provided a demonstration of the dependency of task complexity on the appearance of laterality effects; the increases in task complexity might trigger an increase in the number of engaged biomechanical degrees of freedom during performance. Bryden et al. varied the complexity of the grooved pegboard task. In the standard, more complex version, participants have to pick up each of 25 pegs from a large well, move each toward a peghole, rotate each peg so its orientation matches that of the target peghole, and then place each peg in the hole. In a less complex version of the task, participants have to remove each peg from its peghole and place it in the large well. When dominant and non-dominant hand performance was compared on the two versions of the task, Bryden et al. found no laterality effect in the less complex condition (dominant MD = non-dominant MD); however, dominant-hand superiority was found in the more complex condition (dominant MD < non-dominant MD). Thus, laterality effects are reduced as task complexity declines. Along with variations in task complexity in the peghole task should come variations in engagement of the number of biomechanical degrees of freedom during manipulation of the pegs. Again, the more complex version of the task—but not the easy version—would require rotation of each peg to the orientation of its peghole. Such an increase in task complexity should require more complex engagement of a greater number of joints spanning a larger portion of the upper extremity. Under those conditions, the dominant hand motor control system may be more effective in harnassing those degrees of freedom, which may have given rise to the superior performance of the dominant limb.

In the current study, participants generated cyclical index finger movements involving a single biomechanical degree of freedom, i.e., the finger was splinted and movement was restricted to just the MCP joint (see Figure 1). Based on the analysis of the literature provided so far, one way of understanding the findings from the research on laterality is that laterality effects emerge when system degrees of freedom increase beyond some threshold. When there are many elements that need to be recruited and coordinated in the service of an outcome, the dominant effector and its more elaborate neural circuitry (e.g., Amunts et al., 1996; Sörös et al., 1999; Jung et al., 2003) have the capacity to effectively coordinate all of those elements. In contrast, the non-dominant effector system may be less effective in its capacity to coordinate multiple elements. On the other hand, when a task is of low complexity and the involvement of few degrees of freedom are required, equally effective control of the action is within the capacity of both the dominant and non-dominant motor control systems. In the current study, had the finger been unsplinted—permitting participation of a greater number of index finger joints during load transport—then the dominant motor control system's superior ability to engage, coordinate, and control its degrees of freedom might have emerged. Thus, a further investigation of laterality, load, and motor imagery could examine changes in performance as a function of the number of biomechanical degrees of freedom that can be engaged in the action.

Conclusion

In conclusion, the current study makes three contributions: First, as far as the current author is aware, this is the first study to examine the influence of inertial loading on laterality effects in actual and imagined action. Other recent studies have examined laterality effects in actual and imagined movement in the absence of a load manipulation (e.g., Sabaté, González, & Rodríguez, 2004; Stinear, Fleming, & Byblow, 2006; Skoura et al., 2008; Saimpont, Pozzo, & Papaxanthis, 2009; Nakagawa, Aokage, et al., 2011). Second, although the predicted laterality effects were absent, an analysis of those results in light of the relevant literature led to the hypothesis that the probability of the expected laterality effects should increase with increases in the number of biomechanical degrees of freedom engaged in action. In turn, the need to engage more degrees of freedom should be driven by increases in task complexity or task demand. It is generally recognized that having more available degrees of freedom provides greater flexibility for achieving the goals of action (e.g., Latash, 2012), but here it is also suggested that the dominant hand motor control system might be better able to harness its degrees of freedom and that might result in superior performance of the dominant hand. Again, in the current task, movement was restricted to a single biomechanical degree of freedom, which may have masked the dominant finger from expressing its enhanced capacity to coordinate and control its biomechanical degrees of freedom. Third, the current results for both index fingers replicated the results previously reported by Slifkin (2008) for the dominant finger alone. Namely, for both index fingers, actual-imagined MD differences grew as load increased toward each finger's maximum loading capacity.

References

Amunts, K., Schlaug, G., Schleicher, A., Steinmetz, H., Darringhaus, A. & Roland, P. E. et al. (1996). Asymmetry in the human motor cortex and handedness. *Neuroimage*, *4*, 216-222.

Bagesteiro, L. B. & Sainburg, R. L. (2002). Handedness: Dominant arm advantages in control of limb dynamics. *Journal of Neurophysiology*, *88*, 2408-2421.

Bagesteiro, L. B. & Sainburg, R. L. (2003). Nondominant arm advantages in load compensation during rapid elbow movement. *Journal of Neurophysiology*, *90*, 1503-1514.

Bryden, P. J., Roy, E. A., Rohr, L. E. & Egilo, S. (2007). Task demands affect manual asymmetries in pegboard performance. *Laterality*, *12*, 364-377.

Carson, R. G., Elliott, D., Goodman, D., Thyer, L., Chua, R. & Roy, E. A. (1993). The role of impulse variability in manual-aiming asymmetries. *Psychological Research*, *55*, 291-298.

Carson, R. G., Goodman, D., Chua, R. & Elliott, E. A. (1993). Asymmetries in the regulation of visually guided aiming. *Journal of Motor Behavior*, *25*, 21-32.

Cerritelli, B., Maruff, P., Wilson, P. & Currie, J. (2000). The effect of an external load on the force and timing components of mentally represented actions. *Behavioural Brain Research*, *108*, 91-96.

Decety, J., Jeannerod, M. & Prablanc, C. (1989). The timing of mentally represented actions. *Behavioural Brain Research*, *34*, 35-42.

Desmurget M. & Grafton S. (2000). Forward modeling allows feedback control for fast reaching movements. *Trends in Cognitive Science*, *4*, 423–431

Flowers, K. (1975). Handedness and controlled movement. *British Journal of Psychology*, *66*, 39-52.

Freitas, P. B., Krishnan, V. & Jaric, S. (2007). Force coordination in static manipulation tasks: Effects of the change in direction and handedness. *Experimental Brain Research*, *183*, 487-497.

Garner, W. R. (1962). *Uncertainty and structure as psychological concepts.* New York: John Wiley and Sons.

Gentili, R., Cahouet, V., Ballay, Y. & Papaxanthis, C. (2004). Inertial properties of the arm are accurately predicted during motor imagery. *Behavioural Brain Research*, *155*, 231-239.

Henningsen, H., Ende-Henningsen, B. & Gordon, A. M. (1995). Asymmetric control of bilateral isometric finger forces. *Experimental Brain Research*, *105*, 304-311.

Hoffman, E. R. (1997). Movement time of right- and left-handers using their preferred and non-preferred hands. *International Journal of Industrial Ergonomics*, *19*, 49-57.

Incel, N. A., Ceceli, E., Durukan, P. B., Erdem, H. R. & Yorgancioglu, Z. R. (2002). Grip strength: Effects of hand dominance. Singapore Medical Journal, 43, 234-237.

Jeannerod, M. (1994). The representing brain: Neural correlates of motor intention and imagery. *Behavioral and Brain Sciences*, *17*, 187-245.

Jeannerod, M. (2006). Motor cognition: What actions tell the self. New York: Oxford University Press.

Jung, P., Baumgärtner, U., Bauermann, T., Magerl, W., Gawehn, J. & Stoeter, P. et al. (2003). Asymmetry in the human primary somatosensory cortex and handedness. *Neuroimage*, *19*, 913-923.

Kamarul, T., Ahmad, T. S. & Loh, W. Y. C. (2006). Hand grip strength in the adult Malaysian population. *Journal of Orthopaedic Surgery*, *14*, 172-177.

Kunelius, A., Darzins, S., Cromie, J. & Oakman, J. (2007). Development of normative data for hand strength and anthropometric dimensions in a population of automotive workers. *Work*, *28*, 267-278.

Latash, M. L. (2012). The bliss (not the problem) of motor abundance (not redundancy). *Experimental Brain Research*, *217*, 1-5.

Meyer, D. E., Abrams, R. A., Kornblum, S., Wright, C. E. & Smith J. E. K. (1988) Optimality in human motor performance: ideal control of rapid aimed movements. *Psychological Review*, *89*, 449–482

Nakagawa, K., Aokage, Y., Fukuri, T., Kawahara, Y., Hashizume, A., Kurisu, K. & Yuge, L. (2011). Neuromagnetic beta oscillation changes during motor imagery and motor execution of skilled movements. *Neuroreport*, *22*, 217-222.

Oldfield, R. C. (1971). The assessment and analysis of handedness: The Edinburgh inventory. *Neuropsychologia*, *9*, 97-113.

Papaxanthis, C., Schieppati, M., Gentili, R. & Pozzo, T. (2002). Imagined and actual arm movements have similar durations when performed under different conditions of direction and mass. *Experimental Brain Research*, *143*, 447-452.

Peters, M. & Durding, B. (1979). Left-handers and right-handers compared on a motor task. *Journal of Motor Behavior*, *11*, 103-111.

Plamondon, R. & Alimi, A. M. (1997). Speed/accuracy trade-offs in target-directed movements. *Behavioural & Brain Sciences*, *20*, 279-349.

Provins, K. A. & Magliaro, J. (1989). Skill, strength, handedness, and fatigue. *Journal of Motor Behavior*, *21*, 113-121.

Rabbitt, P. M. A. (1979). How old and young subjects monitor and control responses for accuracy and speed. British Journal of Psychology, 70, 305-311.

Roy, E. A. & Elliott, D. (1986). Manual asymmetries in visually directed aiming. *Canadian Journal of Psychology*, *40*, 109-121.

Roy, E. A. & Elliott, D. (1989). Manual asymmetries in aimed movements. *The Quarterly Journal of Experimental Psychology*, *41A*, 501-516.

Roy, E. A., Kalbfleisch, L. & Elliott, D. (1994). Kinematic analyses of manual asymmetries in visual aiming movements. *Brain and Cognition*, *24*, 289-295.

Sabaté, M., González, B. & Rodríguez, M. (2004). Brain lateralization of motor imagery: Motor planning asymmetry as a cause of motor lateralization. *Neuropsychologia*, *42*, 1041-1049.

Saimpont, A., Pozzo, T. & Papaxanthis, C. (2009). Aging affects the mental rotation of left and right hands. *PloS One*, *4*, e6714.

Sainburg, R. L. (2005). Handedness: Differential specializations for control of trajectory and position. *Exercise and Sport Sciences Reviews*, *33*, 206-213.

Schabowsky, C. N., Hidler, J. M. & Lum, P. S. (2007). Greater reliance on impedance control in the nondominant arm compared with the dominant arm when adapting to a novel dynamic environment. *Experimental Brain Research*, *182*, 567-577.

Schmidt, R. A., Zelaznik, H. N., Hawkins, B., Frank, J. S. & Quinn, J. T. (1979). Motor-output variability: A theory for the accuracy of rapid motor acts. *Psychological Review*, *86*, 415-451.

Skoura, X., Personnier, P., Vinter, A., Pozzo, T. & Papaxanthis, C. (2008). Decline in motor prediction in elderly subjects: Right versus left differences in mentally simulated motor actions. *Cortex*, *44*, 1271-1278.

Slifkin, A. B. (2008). High loads induce differences between actual and imagined movement duration. *Experimental Brain Research*, *185*, 297–307.

Sörös, P., Knecht, S., Imai, T., Gürtler, S., Lütkenhöner, B. & Ringelstein, E. B. et al. (1999). Cortical asymmetries of the human somatosensory hand representation in right- and left-handers. *Neuroscience Letters*, *271*, 89–92.

Stinear, C. M., Fleming, M. K. & Byblow, W. D. (2006). Lateralization of unimanual and bimanual motor imagery. *Brain Research*, *1095*, 139-147.

Stoll, T., Huber, E., Seifert, B., Michel, B. A. & Stucki, G. (2000). Maximum isometric strength: Normative values and gender-specific relation to age. *Clinical Rheumatology*, *19*, 105-113.

Welford, A. T. (1984). Psychomotor performance. In C. Eisendorf (Ed.), *Annual review of gerontology and geriatrics* (237-273). New York: Springer.

Wolpert, D. M., Ghahramani, Z. & Jordan, M. I. (1995). An internal model for sensorimotor integration. *Science*, *269*, 1880–1882.

Wolpert, D. M. & Ghahramani, Z. (2000). Computational principles of movement science. *Nature Neuroscience*, *3*, 1212–1217.

Woodworth, R. S. (1899). The accuracy of voluntary movement. *Psychological Review*, *3* (2, Whole No. 13), 1-114.

In: Motor Behavior and Control: New Research
Editors: Marco Leitner and Manuel Fuchs
ISBN: 978-1-62808-142-8

Chapter 5

Motor Control and Impulsivity: Dysfunctional and Functional Behaviors

Guilherme Menezes Lage,[1,*] Rodolfo Novellino Benda,[1] Anne Marie Mader de Oliveira,[1] Herbert Ugrinowitsch[1] and Leandro Fernandes Malloy-Diniz[2]

[1]Escola de Educação Física, Fisioterapia e Terapia Ocupacional da Universidade Federal de Minas Gerais, Brasil
[2]Faculdade de Medicina da Universidade Federal de Minas Gerais, Brasil

Abstract

Present investigations have demonstrated Motor Control as a possible field for evaluating the adaptive properties of impulsivity. In this chapter we present (a) some possible definitions of impulsivity, (b) the negative impact of impulsivity on motor control, (c) evidences indicating the existence of functional properties of impulsivity, and (d) the first findings indicating an adaptive, functional role of impulsivity on motor control.

Although generally viewed as counterproductive behavior, it seems that impulsivity has a positive impact on motor control in some specific circumstances, in a fast-paced context that requires quick decision-making and fast movements, implicit/automatic processing observed in high-impulsive subjects seems to be productive.

Overall, the relationship between impulsivity and motor control depends on the sensory-motor aspects of the task.

[*] Corresponding author: Guilherme Menezes Lage. Address: EEFFTO_Universidade Federal de Minas Gerais, Av. Presidente Carlos Luz, 6627 - Pampulha - Belo Horizonte – MG; CEP: 31270-901; Phone: 55 31 3409-2394; E-mail: menezeslage@gmail.com.

Introduction

Impulsivity: Definition and Characterization

Impulsivity is a behavioral pattern characterized by several types of manifestations, such as the inability to inhibit an activated or pre-cued response (Dannon et al., 2010), the production of rapid responses without adequate thought, the tendency to act with less forethought and planning, risk-taking behavior motivated by novelty or sensation-seeking tendencies, and mistakes due to lack of focus on the task at hand (Moeller et al., 2001).

Impulsiveness is a core symptom in a large number of psychiatric disorders, such as attention deficit and hyperactivity disorders (Malloy-Diniz et al., 2007), bipolar disorder (Malloy-Diniz et al., 2008), obsessive-compulsive disorder (Da Rocha et al., 2008). However, since subjects who do not meet the criteria for any psychiatric disorder can also present the different dimensions of impulsive behavior, it is the magnitude and frequency of impairments due to impulsiveness that determine its normal or pathological status.[1]

Impulsiveness is not a unitary concept, but rather presents several independent dimensions. For example, Barratt and colleagues (Patton, Stanford, and Barrat, 1995) proposed the existence of three relatively independent dimensions of impulsivity (Figure 1). The motor impulsivity reflects an inhibitory dyscontrol, in which more impulsive subjects tend to act without thinking. High-impulsive subjects demonstrate a difficulty in suppressing inappropriate responses compared to their low-impulsive counterparts (Enticott, Ogloff, and Bradshaw, 2006). The attentional impulsivity denotes lack of focus on the task at hand, in which high-impulsive subjects present difficulty to ignore irrelevant information or thoughts in working memory compared to the low-impulsive subjects. Finally, the non-planning impulsivity is related to the focus on the present without accounting for the consequences of the future outcomes. This dimension is related to the intolerance to delay gratification, materialized by an increased preference for immediate reward over more advantageous but delayed reward (Pattij and Vandershuren, 2008). An example occurs in gambling tasks, where high risk strategy to obtain immediate reward is maintained by more impulsives regardless of the possible negative consequences of this approach. The non-planning impulsivity is also known as impulsive decision making (Pattij and Vandershuren, 2008) or cognitive impulsivity (Bechara, Damasio and Damasio, 2000).

The main instrument to evaluate these dimensions of impulsivity is the BIS-11 scale, a self-report questionnaire with 30 items to score the level of motor, attentional, non-planning impulsivity.

Recently, computerized tasks to assess the different mechanisms of impulse control have also been used. The continuous performance test (CPT-II, Epstein et al., 2003) is a common test used to measure motor and attentional impulsivity (Lage et al., 2011a; Malloy-Diniz et al., 2007; Swann et al., 2002; Walderhaug et al., 2007; Walderhaug, Landrø and Magnusson, 2008). Another task is the Iowa Gambling Task (IGT). Maintenance of a high-risk strategy in the IGT reflects sustained engagement of a particular behavior despite ongoing evidence that it is dysfunctional.

[1] In this chapter, we described essentially the impulsive behavior of the non-clinical population. The comparison between high- and low-impulsive subjects is not related to the pathological aspects, but refers to the normal and expected differences in personality traits among people.

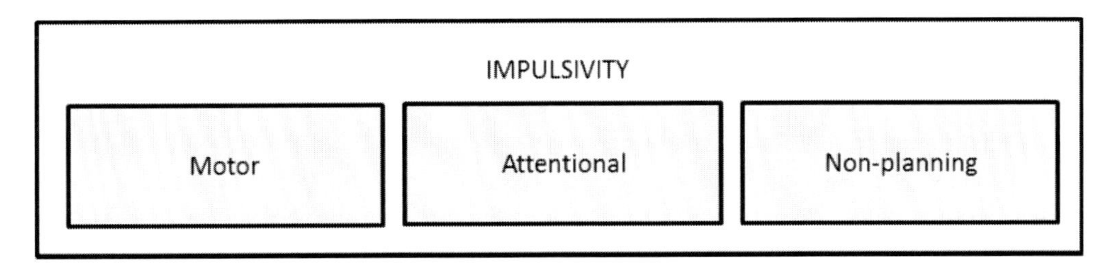

Figure 1. Impulsiveness as a multidimensional construct, an example is the Barratt model (Patton, Stanford, and Barrat, 1995).

The IGT models real-life decision making, especially the types of decisions that are consistent with the construct of non-planning (cognitive) impulsivity (Bechara and Van Der Linden, 2005; Malloy-Diniz et al., 2007, Malloy-Diniz et al., 2008).

The Barratt multidimensional model (Patton et al., 1995) has been supported by several studies derived from both clinical (e.g., Salgado et al., 2008) and normal samples (e.g., Spinella, 2005), and has been correlated with neuropsychological measures (Malloy-Diniz et al., 2007). Different neurobiological substrates seem to be associated with each one of the three dimensions of impulsivity. Patients with ventromedial prefrontal cortex damage do not present high levels of motor impulsivity, although being markedly unable to delay gratification of reward (Bechara, Damasio and Damasio, 2000). Motor impulsivity is associated with the more posterior regions of the orbitofrontal/ventromedial prefrontal cortex, including the basal forebrain and dorsolateral prefrontal cortex, while non-planning impulsivity is associated with the more anterior orbitofrontal/ventromedial prefrontal cortex, including the frontal pole. The inability to inhibit irrelevant information held in working memory and to focus on the task at hand is linked to the dorsolateral prefrontal cortex (Bechara and Van Der Linden, 2005) and anterior cingulate (Malloy-Diniz et al., 2008).

Association between impulsivity and neurobiological substrates has been found not only in anatomo-functional features, but also at the molecular level (Lage et al., 2011a). Dopaminergic and serotonergic modulation has an important impact on the impulsive phenotype. The hypofunction of the dopaminergic system has been associated with motor impulsivity (Congdon, Lesch and Canli, 2008) and non-planning impulsivity (Limosin et al., 2003). The same is true to the serotonergic system, low levels of serotonin are associated with attentional impulsivity (Sakado et al., 2003).

The impulsive trait is explained in part by the genotype. In particular, there is support for two polymorphisms in genes of the dopaminergic system, the dopamine D4 receptor (DRD4) and the dopamine transporter (DAT), as well as one gene of the serotonergic system, the serotonin transporter (5-HTT). The gene coding for the D4 receptor contains a variable number of tandem repeats (VNTR) polymorphism that varies from 2 to 11 repeats across individuals. Presence of the 7-repeat allele has been associated with motor impulsivity (Congdon, Lesch and Canli, 2008). The gene coding for the protein that controls the uptake of dopamine from the synaptic cleft, the DAT, also contains a VNTR polymorphic region that varies between individuals. The 10/10 DAT genotype has been associated with a measure of behavioral inhibition (Cornish et al., 2005). The serotonin transporter gene influences the magnitude and duration of serotonergic activity by controlling the uptake of serotonin from the extracellular space.

This gene has a functional polymorphism (5-HTTLPR) in its regulatory region that regulates the transcription of the 5-HTT and has been associated with motor impulsivity (Walderhaug et al., 2010) and attentional impulsivity (Sakado et al., 2003).

It is important to note that impulsive behavior seems to be a component of human nature since it appears in almost every one and its expression varies along lifespan (de Luca and Leventer, 2008). The persistence of impulsive behavior in our species occurs due its beneficial consequences considering both positive and negative reinforcement in settings where fast decisions are required (Dickman, 1990). Nonetheless, in some cases, impulsive behavior is over expressed and leads to constant impairments as in neurological and psychiatric disorders. Therefore, it is important to study impulsivity both in normal and pathological settings to understand its different types of manifestation.

Impulsivity and Motor Control

Several studies about impulsivity have been conducted in different populations and in different domains of human behavior, such as the cognitive and the social, but in comparison, little has been investigated in the motor domain.

The studies that compared the motor performance of high- and low-impulsive subjects can be distinguished by the features of the task employed. Some studies used discrete tasks with low demand on the motor system. These tasks are characterized by pressing one or more keys to a stimuli appearance, in which the primary variable evaluated was the reaction time (Cohen and Horn, 1974; Dickman and Meyer, 1988; Expósito and Andrés-Pueyo, 1997; Logan, Schachar and Tannock, 1997; Rodriguez-Fornells, Lorenzo-Seva and Andrés-Pueyo, 2002). Other studies adopted tasks with different demands. Bachorowski and Newman (1985; 1990) also applied a discrete task, but the subjects were asked to trace a circle as slow as possible. Some studies applied cyclical, repetitive tasks also with low demand to the motor system, such as finger tapping with requirement of tapping as quickly as possible (Amelang and Breit, 1983; Matthews and Jones; Chamberlain, 1989) or of timing production (Barrat, 1981). In the following studies (Amelang and Breit, 1983; Bachorowski and Newman, 1985; 1990; Barrat, 1981; Matthews and Jones; Chamberlain, 1989), the main variable evaluated was the movement time. Pursuit rotor task requires the subject keep a stylus on a mark on a rotating metal disk, where the primary variable is the accuracy (Barrat, 1967; Smith et al., 1991). Finally, aiming tasks have been used in some studies (Lage et al., 2012a; Lemke, 2005; Lemke et al., 2005) and performance variables such as reaction and movement time have been used in association with kinematics variables such as peak velocity and discontinuities in acceleration. The use of kinematic analysis is important since the association between impulsivity and motor control is better explained in terms of mechanisms of control.

Distinctions between less and more impulsives have been found in these studies, indicating differences in reaction time tasks (Expósito and Andrés-Pueyo, 1997), tapping at a specified rate (Barrat, 1981), pursuit rotor (Barrat, 1967), drawing (Bachorowski and Newman, 1985, 1990) and aiming (Lage et al., 2012a; Lemke, 2005; Lemke et al., 2005). In spite of some controversial findings (cf. Dickman, 1993), the general pattern over the course of these studies has indicated that impulsivity is related to faster but less accurate movements (Lage et al., 2012a).

To our present knowledge, only three studies used techniques of motion analysis to assess possible differences between less and more impulsives in terms of kinematic measures (Lage et al., 2012a; Lemke, 2005; Lemke et al., 2005).

In these studies, kinematical analysis was used to investigate the role of impulsivity in the control of goal-directed aiming tasks. Lemke and colleagues (Lemke, 2005; Lemke et al., 2005) observed that healthy individuals with higher impulsivity scores on BIS-11 presented shorter relative time to achieve the peak velocity than the less impulsives.

There were no differences between less and more impulsives in movement time, peak velocity and movement trajectory. Lage et al. (2012a) found similar results in relation to the relative time to achieve the peak velocity. It shows differences in the planning and execution of the aiming movements between low- and high-impulsive subjects. The time interval preceding the peak velocity, which is the initial impulse, reflects the preprogrammed characteristics of the movement (Lage et al., 2012b).

After the peak velocity was achieved, an error correction phase or final homing component occurs (Khan et al., 2006). The lower duration of the initial impulse, as demonstrated by the high-impulsive subjects, was reflective of a motor control strategy that was based more on a closed-loop control, that is, a type of control that relies more on visual feedback to produce online corrections. In contrast, the low-impulsive subjects preprogrammed their movements to achieve a longer initial impulse to home, nearest to the target, resulting in minimum error correction (Figure 2).

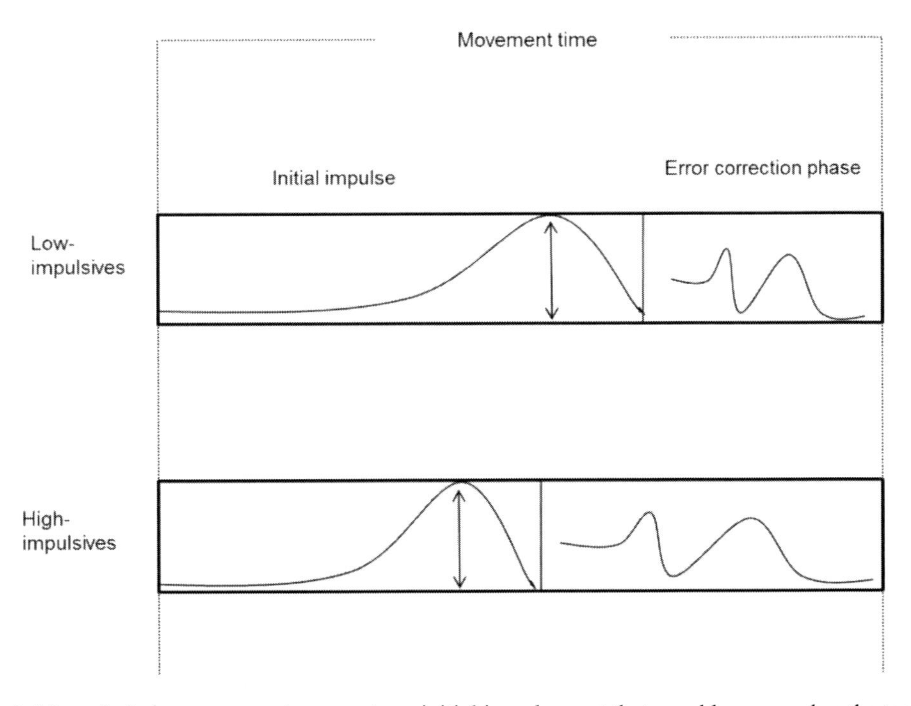

Figure 2. Manual aiming movements presents an initial impulse part that roughly approaches the target by open-loop control, and a final homing part under closed-loop control, with adjustments visually guided in the last portion of the movement. A usual kinematic marker used to distinguish the two components of the movement is the peak velocity. High-impulsive subjects rely more on closed-loop control than the low-impulsive subjects.

A discrepancy in the results of Lemke et al. (2005) and Lage et al. (2012a) was found. There was no difference between less and more impulsives in peak velocity in the Lemke's study. In contrast, Lage et al. (2012a) observed that high-impulsive subjects achieved higher levels of peak velocity during their arm trajectories than the low-impulsive subjects. Different from Lemke et al. (2005), who did not require fast aiming movements, Lage et al. (2012a) instructed the participants to execute the movement to the target as quickly as possible. These results show that sensory-motor aspects of the task interfere in the relationship between motor control and impulsivity.

Interrelations between neural regions may be involved in the differences observed in motor performance between less and more impulsives. Increasing evidence suggests the participation of specific frontal lobe areas in motor behavior (Serrien, Ivry and Swinnen, 2007) indicating that the psychophysiological aspects involved in impulsivity may interfere on motor control.

Cognitive resources are recruited to ensure the successful realization of the action goal and a key region in this process is the dorsolateral prefrontal cortex. The dorsolateral prefrontal cortex enable us to hold information in mind to remember the supposed goal, to avoid distraction and stay on task, to resist to responding too early, and to inhibit a prepotent response (Diamond, 2000).

All these functions are related to the response selection and monitoring. From the three dimensions of impulsivity, only motor and attentional impulsivities are associated with this neural region.

In a behavioral study, Lage et al. (2012a) confirmed their hypothesis that motor impulsivity would be more related to motor control than non-planning impulsivity. Anatomically speaking, the dorsolateral prefrontal cortex has extensive interconnections with regions directly involved in motor functions (Figure 3), such as premotor cortex and the supplementary motor area (Dum and Strick, 1991; Tanji, 1994).

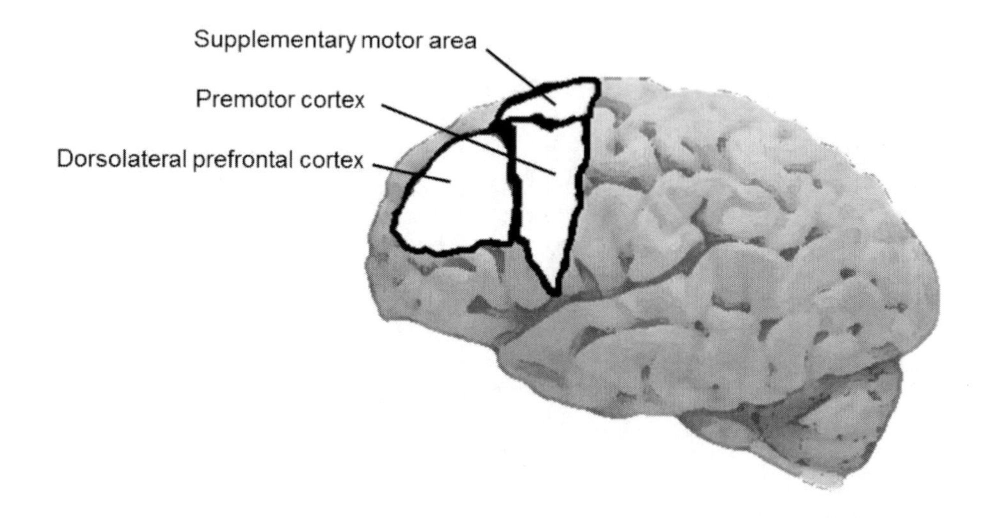

Figure 3. Neurobiological interactions between impulsivity and motor control. Interconnections are observed among a critical area for several complex cognitive functions, the dorsolateral prefrontal cortex, with motor areas, the premotor cortex and supplementary motor area.

The Adaptive and Maladaptive Dimensions of Impulsivity

The majority of research efforts are focused on the analysis of the dysfunctional properties of impulsivity, independent of the context or human domain investigated. Usually, when fast, automatic, responses produces positives outcomes, people tend to classify these behaviors not as impulsivity, but as an indicative sign of spontaneity, rapidity, courage or creativity (Eysenck, 1993). An exception observed in the literature is the model proposed by Dickman (1990), in which two relatively independent types of impulsivity co-exist when analyzing the subject's personality. On the first, dysfunctional impulsivity is associated with the classic concept of inhibitory dyscontrol underlying the impulsive behavior. On the second, the concept of functional impulsivity is related to the characteristic of the subject to think, act and speak quickly without consistent lost in accuracy. Subjects with high levels of functional impulsivity tend to emphasize speed over certainty of accuracy when the situation requires it (Reeve, 2007).

According to Dickman (1990, p.98), "functional impulsivity was more closely associated with enthusiasm (i.e., rhathymia), adventurousness, and activity than was dysfunctional impulsivity". The association between functional impulsivity and these traits seem to explain why subjects with high levels of functional impulsivity report more frequently the benefits from their impulsivity than the dysfunctional impulsives subjects. "Enthusiastic, active individuals who are willing to take risks are likely to be very productive; in such individuals, the sheer quantity of their output could well compensate for the high number of errors in that output" (Dickman, 1990, p.98). In contrast, subjects with high levels of dysfunctional impulsivity display disorderliness and tendency to ignore relevant facts when making decisions.

Since the Dickman (1990) model describes functional and dysfunctional impulsivity as independent dimensions, it is possible to observe a person with high (or low) on both dimensions; being high on one dimension does not imply the subject is low on the other (Reeve, 2007).

For example, an athlete high on both functional and dysfunctional impulsivity perform well in a fast-paced context where quick decision making and fast movements are required, but would demonstrate difficulty to engage in slow, deliberate processing when the circumstance requires this type of behavior.

Conversely, an athlete with similar skills, but high on functional impulsivity and low on dysfunctional impulsivity, might perform well in both fast- and low-paced contexts showing an amazing adaptive ability and feeling comfortable in different situations of performance (Table 1). Dickman (1990) developed a 23 items self-reported questionnaire with two sub-scales to measure the tendency to act with relatively little forethought when this causes problems (dysfunctional impulsivity) and when this is optimal (functional impulsivity).

Studies that applied tasks of cognitive functioning have found the adaptive facet of impulsivity. For example, when the experimental task is very simple, high-impulsive subjects produce rapid responses with few errors (Dickman, 1985).

Moreover, when the time available for making a decision is very constrained, high-impulsive subjects are in fact more accurate than low-impulsive subjects (Dickman and Meyer, 1988).

Table 1. Expected performance when combining functional and dysfunctional traits with fast-and low-paced contexts

Context	Functional and dysfunctional traits			
	High/low	High/high	Low/low	Low/high
	Performance	Performance	Performance	Performance
Fast-paced	High	High	Low	Low
Low-paced	High	Low	High	Low

It is important to highlight that these studies did not use the Barrat model (tridimensional), as well as the Dickman model (bidimensional) as theoretical frameworks to explain the findings. The motor domain seems to be an important field to evaluate the possible adaptive properties of impulsivity. In several contexts, the temporal and/or spatial demand to the motor system is high, requiring different levels of trade-off between velocity and accuracy. Thus, rapid, error-prone information processing is an optimal strategy in some specific circumstances. Here, it is important to emphasize that dysfunctional impulsivity is related to the engagement in rapid, error-prone information processing because of an inability to use a slower, more methodical approach under certain conditions (Dickman, 1990).

Functional Impulsivity and Motor Control

To our knowledge, there are no specific studies that investigated the impact of the functional impulsivity on motor control, independently of the possible theoretical framework used. However, some inferences about this relationship are possible when analyzing some specific results from recent studies.

Lage et al. (2012a) employed an aiming task, in a digitizing tablet, in which the participants were required to make fast and accurate strokes with a pen from the home position to the target; these strokes were displayed in real time on a computer monitor. The aiming task presented four conditions of execution that required different perceptual-motor demands. The control condition was the prepotent condition, that is, a circumstance that is frequently required during the task execution. The distractor condition epitomized a situation in which irrelevant information, related to the prepotent condition, needed to be ignored for a successful goal achievement. The response inhibition condition represented an adaptive context characterized by the necessity of suppressing a prepotent response.

Finally, the higher index of difficulty condition symbolizes an adaptive context in which a more difficult situation, relative to the control condition, is set to the performer.

Lage et al. (2012a) tried to recreate some perceptual requirements found in Go/No-Go neuropsychological tasks (Epstein et al., 2003; Walderhaug et al., 2007) but, that at the same time, demanded more of the motor system than just pressing a button with a finger.

The main point in discussing the role of functional impulsivity in motor control is based on the comparison between high- and low-impulsive subjects in the control condition and the higher index of difficulty condition. We can assume that control condition was the low-paced condition and the higher index of difficulty condition was the fast-paced condition.

Instead of the maximum time to achieve the target after the appearance of the imperative stimuli being the same for both conditions (2 sec), two important differences are clear in these two execution conditions: (a) control condition appeared in 70% of the trials, while the higher index of difficulty condition appeared in 10% of the trials (the other 20% of trials appeared in the distractor condition and in inhibition of response condition); (b) index of difficulty (ID) in the control condition was 5.2 bits (Fitts, 1954), while in the higher index of difficulty condition the ID was 6.3 bits. In other words, the uncertainty of stimuli appearance and the spatial and temporal demand in the higher index of difficulty condition were higher when compared to the control condition.

The results showed that in the condition with higher demand to the motor system, the higher index of difficulty condition, high-impulsive participants exhibited a lower frequency of incorrect hits to the target than the low-impulsive subjects. It was expected that low-impulsive subjects exhibit a greater spatial accuracy than their more impulsive counterparts. Lage et al. (2012a) proposed that the relationship between impulsivity and motor control depends on the sensory-motor aspects of the task and that in situations in which the temporal and spatial demands to the motor system are high, impulsivity has a functional, adaptive effect on the behavior.

The explanation to this finding is based on the differences between the types of processing information observed in low- and high-impulsive subjects. Information processing can be explicit, characterized by conscious, controlled and reflective processes, or implicit, ruled by unconscious, automatic and intuitive processes.

In controlled processing, behavior is assumed to be a consequence of a decision process; in impulsive/automatic processing, behavior is a consequence of automatically spreading activation in an associative network (Richetin and Richardson, 2008). In the higher index of difficulty condition, low-impulsive participants did not achieve, in several trials, the small target (0.5 cm in diameter) in the maximum time of the task (2 sec). This means the task closed before the cursor had achieved the target. The difficulty of the low-impulsive subjects in this context was not related to the degree of accuracy, but to the slowness to prepare and execute the movement.

Although generally viewed as counterproductive behavior (Stanford et al., 2009), it seems that impulsivity has a positive impact on motor control in some specific circumstances. These results reinforce the idea that the performance of high- and low-impulsive subjects can be favored or not favored according to the context. In a fast-paced context where quick decision making and fast movements are required, implicit/automatic processing seems to be productive.

Another example is found in Lage et al. (2011b). They investigated the relationship between impulsivity and technical performance in female handball athletes. Their main justification to investigate this relationship was that sportive contexts are a complex and dynamic environment where impulsiveness probably interferes in the motor control. It is true mainly in "... open-skill sports (e.g., soccer, basketball and handball). Due to constant changes in the environment (e.g., alterations in opponents positioning), the player is forced to inhibit pre-planned responses, anticipate actions and coordinate corporal segments based on the complex and dynamic flow of sensorial information" (Lage et al., 2011b, p.721).

In this study (Lage et al., 2011b) non-planning impulsivity was positively correlated with some measures of technical faults, as for example carrying the ball for more than three steps.

At first sight, this finding seems paradoxical because low non-planning impulsivity was correlated with a higher occurrence of this technical fault. Open-skill sports are characterized by a high level of uncertainty that forces the player to use complex cognitive skills under temporal pressure (Ripoll, Kerlirzin, Stein, and Reine, 1995), but decision-making involve slow, conscious and effortful reflections about possible consequences (Bechara and Van der Linden, 2005). Hence, it is possible that players with low non-planning impulsivity emphasize the accuracy of motor responses, thereby causing a temporal cost in situations in which the speed of information processing is an essential feature. In contrast, players with high non-planning impulsivity would have performed better in match situations in which they needed to think and respond quickly. One could ask whether positive correlations between impulsivity and motor performance would also be found in a task that does not require fast responses. However, in a third study, Lage and colleagues (Lage et al., 2011b) did not find indications of functional impulsivity in a slow-pace task (timing motor task), where they found only the presence of dysfunctional impulsivity.

Conclusion

Overall, these findings of studies indicate that the psychophysiological aspects involved in impulsivity interfere on motor control. The interconnections of the dorsolateral prefrontal cortex with regions directly involved in motor functions such as premotor cortex and the supplementary motor area seems to be a neurobiological substrate associated with the behavioral findings. The impact of dysfunctional impulsivity on motor control is clear, but some initial findings show us a promising and exciting field of research. It is possible that in a near future we will also recognize impulsivity, at least in specific motor contexts, as a positive trait.

References

Amelang, M., Breit, C. (1983). Extraversion and rapid tapping: reactive inhibition or general cortical activation as determinants of performance differences. *Personality and Individual Differences*, 4, 103-105.

Bachorowski, J. A. and Newman, J. E. (1985). Impulsivity in adults: Motor inhibition and time estimation. *Personality and Individual Differences*, 6, 133-136.

Bachorowski, J. A. and Newman, J. E. (1990). Impulsive Motor Behavior: Effects of Personality and Goal Salience. *Journal of Personality and Social Psychology*. 58, 512-518.

Barrat, E. S. (1981). Time perception, cortical evoked potentials and impulsiveness among three groups of adolescents. In: J. K. Hays, T. K. Roberts and K. S. Solway (Eds.), *Violence and the violent individual* (pp.87-95). New York: Spectrum Publications.

Barrat, E. S. (1967). Perceptual-motor performance relative to impulsivity and anxiety. *Perceptual and Motor Skills*, 25, 485-492.

Bechara, A. and Van Der Linden, M. (2005). Decision-making and impulse control after frontal lobe injuries. *Current Opinion in Neurology*, 18, 734-739.

Bechara, A., Damasio, H. and Damasio, A. R. (2000). Emotion, decision making and the orbitofrontal cortex. *Cerebral Cortex*, 10, 295-307.

Cohen, D. B. and Horn, J. M. (1974). Extraversion and performance: a test of the theory of cortical inhibition. *Journal of Abnormal Psychology*, 83, 304-307.

Congdon, E., Lesch, K. P. and Canli, T. (2008). Analysis of DRD4 and DTA polymorphisms and behavioral inhibition in health adults: implications for impulsivity. *American Journal of Medical Genetics Part B: Neuropsychiatry genetics*, 147b, 27-32.

Cornish, K. M., Manly, T., Savage, R., Swanson, J., et al. (2005). Association of the dopamine transporter (DAT1) 10/10-repeat genotype with ADHD symptoms and response inhibition in a general population sample. *Molecular Psychiatry*, 10, 686–698.

Da Rocha F., Malloy-Diniz, L. F., DeSouza, K., Correa, H., and Teixeira, A. (2008). Borderline personality features possibly related to cingulate and orbitofrontal cortices dysfunction due to schizencephaly. *Clinical Neurology and Neurosurgery*, 110, 396-399.

Dannon, P. N., Shoenfeld, N., Rosenberg, O., Kertzman, S., and Kotler, M. (2010). Pathological gambling: An impulse control disorder? Measurement of impulsivity using neurocognitive tests. *Israel*, 12, 243-248.

De Luca, C. R. and Leventer, R. J. (2008). Developmental trajectories of executive functions across the lifespan. In: Anderson, P., Anderson, V. and Jacobs, R. *Executive functions and the frontal lobes: a lifespan perspective*. Washington, DC: Taylor and Francis. pp.3–21.

Diamond, A. (2000). Close Interrelation of Motor Development and Cognitive Development and of the Cerebellum and Prefrontal Cortex. *Child Development*, 71, 44-56.

Dickman, S. (1985). Impulsivity and perception: Individual differences in the processing of the local and global dimensions of stimuli. *Journal of Personality and Social Psychology*, 48, 133-149.

Dickman, S. (1990). Functional and dysfunctional impulsivity: personality and cognitive correlates. *Journal of Personality and Social Psychology*, 58, 95-102.

Dickman, S. J. and Meyer, D. E. (1988). Impulsivity and speed-accuracy tradeoffs in information processing. *Journal of Personality and Social Psychology*, 54, 274-290.

Dum, R. P. and Strick, P. L. (1991). The origin of corticospinal projections from the premotor areas in the frontal lobe. *Journal of Neuroscience,* 11, 667-689.

Enticott, P. G., Ogloff, R. P. and Bradshaw, J. (2006). Associations between laboratory measures of executive inhibitory control and self-reported impulsivity. *Personality and Individual Differences*, 41, 285-294.

Epstein, J. N., Erkanli, A., Conners, C. K., Klaric, J., Costello, J. E., and Angold, A. (2003). Relations between continuous performance test performance measures and ADHD behaviors. *Journal of Abnormal Child Psychology*, 31, 543-554.

Expósito, J. and Andrés-Pueyo, A. (1997). The effects of impulsivity on the perceptual and decision stages in a choice reaction time task. *Personality and Individual Differences*, 22, 693-697.

Eysenck, H. J. (1993). The nature of impulsivity. In: Mccown, W., Shure, M. and Johnson, J. (Eds.), *The impulsive client: theory, research and treatment*. Washington, DC: American Psychological Association. pp.57-69.

Fitts, P. M. (1954). The information capacity of the human motor system in controlling the amplitude of movement. *Journal of Experimental Psychology*, 47, 381-391.

Khan, M. A., Franks, I. M., Elliott, D., Lawrence, G. P., Chua, R., Bernier, P. M., Hansen, S., and Weeks, D. J. (2006). Inferring online and offline processing of visual feedback in target-directed movements from kinematic data. *Neuroscience and Biobehavioral Reviews*, 30, 1106-1121.

Lage, G. M., Malloy-Diniz, L. F., Matos, L. O., Bastos, M. A., Abrantes, S. C., and Correa, H. (2011a). Impulsivity and the 5-HTTLPR polymorphism in a non-clinical sample. *PLoS ONE*, 6, e16927.

Lage, G. M., Malloy-Diniz, L. F., Fialho, J. V. A., Gomes, C. M., Albuquerque, M. R., and Corrêa, H. (2011b). Correlação entre as dimensões da impulsividade e o controle em uma tarefa motora de timing [Correlation between the dimensions of the impulsivity and the control of a timing motor task]. *Brazilian Journal of Motor Behavior*, 6, 39-46.

Lage, G. M., Malloy-Diniz, L. F., Moraes, P. H. P., Neves, F. S., and Corrêa, H. (2012a). A kinematic analysis of the association between impulsivity and manual aiming control. *Human Movement Science*, 31, 811-823.

Lage, G. M., Malloy-Diniz, L. F., Neves S. F., Gallo, L. G., Valentini, A. S., Corrêa, H. (2012b). A kinematic analysis of manual aiming control on euthymic bipolar disorder. *Psychiatry Research*, doi:10.1016/j.psychres.2012.09.046.

Lemke, M. R. (2005). Impulsivität und MotoriK. *Psychoneuro*, 31, 385-387.

Lemke, M. R., Fischer, C. J., Wendorff, T., Fritzer, G., Rupp, Z., and Tetzlaff, S. (2005). Modulation of involuntary and voluntary behavior following emotional stimuli in healthy subjects. *Progress in Neuro-Psychopharmacology and Biological Psychiatry*, 29, 69-76.

Limosin, F., Loze, J. Y., Dubertret, C., Gouya, L., Ades, J., Rouillon, F., and Gorwood, P. (2003). Impulsiveness as the intermediate link between the dopamine receptor D2 gene and alcohol dependence. *Psychiatry Genetics,* 13, 127-129.

Logan, G. D., Schachar, R. J., Tannock, R. (1997). Impulsivity and inhibitory control. *Psychological Science*, 8, 60-64.

Malloy-Diniz, L. F., Fuentes, D., Leite, W. B. Correa, H., and Bechara, A. (2007). Impulsive Behavior in Adults with ADHD: characterization of motor, attentional and cognitive impulsiveness. *Journal of the International Neuropsychological Society*, 13, 693-698.

Malloy-Diniz, L. F., Leite, W. B., DeMoraes, P. H. P., Correa, H., Bechara, A., and Fuentes, D. (2008). Brazilian Portuguese version of the Iowa Gambling Task: transcultural adaptation and discriminant validity. *Revista Brasileira de Psiquiatra*, 30, 144-148.

Matthews, G., Jones, D. M., Chamberlain, A. G. (1989). Interactive effects of extraversion and arousal on attentional task performance: multiple resources or enconding processes? *Journal of Personality and Social Psychology*, 56, 629-639.

Moeller, F. G., Barrat, E. S., Dougherty, D. M., Schmitz, J. M., and Swann, A. C. (2001). Psychiatry aspects of impulsivity. *American Journal of Psychiatry*, 158, 1783-1793.

Pattij, T. and Vandershuren, L. J. M. J. (2008). The neuropharmacology of impulsive behavior. *Trends in Pharmacological Sciences*, 29, 192-199.

Patton, J. H., Stanford, M. S., Barrat, E. S. (1995). Factor structure of the Barratt impulsiveness scale. *Journal of Clinical Psychology*, 51, 768-774.

Reeve, C. L. (2007). Functional Impulsivity and Speeded Ability Test Performance. *International Journal of Selection and Assessment*, 15, 56-62.

Richetin, J. and Richardson, D. S. (2008). Automatic processes and individual differences in aggressive behavior. *Aggression and Violent Behavior*, 13, 423–430.

Ripoll, H., Kerlirzin, Y., Stein, J., and Reine, B. (1995). Analysis of information processing decision making, and visual strategies in complex problem solving sport situations. *Human Movement Science*, 14, 325-349.

Rodriguez–Fornells, A., Lorenzo-Seva, U. and Andrés-Pueyo, A. (2002). Are high impulsive and high risk taking people more motor disinhibited in the presence of incentive? *Personality and Individual differences*, 32, 661-683.

Sakado, K., Sakado, M., Mundtm, C., and Someya, T. (2003). A psychometrically derived impulsive trait related to a polymorphism in the serotonin transporter gene-linked polymorphic region (5-HTTLPR) in a Japanese nonclinical population: Assessment by the Barratt Impulsiveness Scale (BIS). *American Journal of Medical Genetics*, 121B: 71-75.

Salgado, J. V., Malloy-Diniz, L. F., Campos, V. R., Abrantes, S. S., and Fuentes, D., et al. (2009). Neuropsychological assessment of impulsive behavior in abstinent alcohol-dependent subjects. *Revista Brasileira de Psiquiatria*, 31, 4-9.

Scrricn, D. J., Ivry, R. B. and Swinnen, S. P. (2007). The missing link between action and cognition. *Progress in Neurobiology*, 82, 95–107.

Smith, A. P., Rusted, J. M., Savory, M., Eaton-Willians, P., and Hall, S. R. (1991). The effects of caffeine, impulsivity and time of day on performance, mood and cardiovascular function. *Journal of Psychopharmacology*, 5, 120-128.

Spinella, M. (2005). Prefrontal substrates of empathy: Psychometric evidence in a community sample. *Biological Psychology*, 70, 175-181.

Stanford, M. S., Mathias, C. W., Dougherty, D. M., Lake, S. L., Anderson, N. E., and Patton, J. H. (2009). Fifty years of the Barratt Impulsiveness Scale: An update and review. *Personality and Individual Differences*, 47, 385-395.

Swann, A. C., Bjork, J. M., Moeller, F. G., and Dougherty, D. M. (2002). Two models of impulsivity: Relationship to personality traits and psychopathology. *Biological Psychiatry,* 51, 988–994.

Tanji, J. (1994). The supplementary motor area in the cerebral cortex. *Neuroscience Research,* 19, 251-268.

Walderhaug, E., Landrø, N. I. and Magnusson, A. (2008). A synergic effect between lowered serotonin and novel situations on impulsivity measured by CPT. *Journal of Clinical and Experimental Neuropsychology*, 30, 204–211.

Walderhaug, E., Magnusson, A., Neumeister, A., Lappalainen, J., Lunde, H., Refsum, H., and Landrø, N. I. (2007). Interactive effects of sex and 5-HTTLPR on mood and impulsivity during tryptophan depletion in healthy people. *Biological Psychiatry*, 62, 593–599.

In: Motor Behavior and Control: New Research
Editors: Marco Leitner and Manuel Fuchs
ISBN: 978-1-62808-142-8

Chapter 6

The Contribution of Better Understanding Martial Arts Strikes to Studies in Motor Control

Osmar Pinto Neto[1,2,*]
[1]Arena235 Consultoria Esportiva - Sao José dos Campos SP, Brazil
[2]Instituto de Engenharia Biomédica / Universidade Camilo Castelo Branco
Sao José dos Campos SP, Brazil

Abstract

Martial Artists distinguish themselves from regular people as they constantly train their bodies to be strong, precise, accurate and fast. However, a increase in one of these characteristics may be detrimental to the others. Additionally, quantifying striking peak force is most often challenging and determining what other physical variables are reliable when trying to understand within-subject and between-subject peak force variations is worthwhile. As such, this study has two main goals. The first goal is to investigate the possible trade off between peak hand acceleration and accuracy and consistency of hand strikes performed by martial artists of different training experiences. The second is to investigate the correlation among several physical variables commonly used to quantify the performance of martial arts and combative sports strikes. Thirteen martial artists (10 male and 3 female) volunteered to participate in the experiment. Each participant performed 12 maximum effort goal-directed strikes targeted at an instrumented pendulum. The target was instrumented with one load cell, a pressure sensor, and a tri-axial accelerometer block. Additionally, hand acceleration during the strikes was obtained using tri-axial accelerometer block. We estimated subject's accuracy, precision, hand's speeds before impact, peak accelerations before and during impact, the strike's peak force, pendulum's peak acceleration. We found that for our male subjects training experience was significantly correlated to hand peak acceleration prior to impact (r^2 = 0.456, $p = 0.032$) and accuracy (r^2 = 0. 621, $p = 0.012$). These correlations suggest that more experienced participants exhibited higher hand peak accelerations and at the same

[*] Corresponding author: E-mail: osmarpintoneto@hotmail.com.

time were more accurate. Training experience, however, was not correlated to consistency ($r^2 = 0.085$, $p = 0.413$). Overall, these results suggest that martial arts training may lead practitioners to achieve higher striking hand accelerations with better accuracy and no change in striking consistency. Furthermore, considering within-subject variations, we found that peak hand acceleration during the impact (mean r^2=0,56) followed by peak hand speed before impact (r^2=0.33) were the variables that exhibited greater correlation to peak force. As for between-subject comparisons, we found that peak hand acceleration before impact (r^2=0.82) and hand speed before impact (r^2=0.82) were highly correlated to peak force.

Keywords: Motor output variability, combative sports; accelerometer, pressure sensor

Introduction

Martial artists distinguish themselves from regular people as they constantly train their bodies to be strong, precise, accurate and fast. However, a increase in one of these characteristics may be detrimental to the others. Additionally, quantifying striking peak force is most often challenging and deciding what other physical variables are correlated to peak force when trying to understand within-subject and between-subject peak force variations is worthwhile.

It has been suggested that in order to improve end-point accuracy in goal-directed tasks the central nervous system (CNS) may select motor patterns which cause a decrease in motor-output variability (Muller & Sternad, 2004). The minimum variance theory, proposed by Harris and Wolpert (1998), suggests that the cause of motor output variability is that neural control signals are corrupted by noise whose magnitude increases with the size of the control signal. Thus, one may expect that with the increase in magnitude of the neural control signals a movement becomes more variable (trajectory and end-point) and less accurate (Harris & Wolpert, 1998; Wolpert & Ghahramani, 2000). By increase in the neural control signal, Harris & Wolpert (1998) meant either an increase in the firing of motor units and/or an increase in the magnitude of the movement (velocity and/or acceleration). Although this theory has been tested with laboratory controlled movements, it has not been tested with sport movements, especially martial arts strikes.

Apart from the original study by Harris & Wolpert, 1998, where they found evidence to support the minimum variance theory using experimental data from Uno et al. (1989), there have been experimental evidence that support (Christou et al., 2003; Muller & Sternad, 2004) and also contradict (Corcos et al., 1993, Christou et al., 2007, Poston et al., 2010) the theory. Christou et al. (2003) analyzed concentric and eccentric contractions with the first dorsal interosseus muscle (FDI) and found that, consistent with the minimum variance theory, movement accuracy was related to the fluctuations in acceleration for both types of contractions. Muller & Sternad (2004) used a simplified virtual version of the British pub game "skittles and showed that variability in performance after practice was reduced in greater part by reduction of stochastic noise. On the other hand, Corcos et al (1993) analyzed rapid elbow flexions movements and found that subjects enhanced performance after practice, increasing movement speed and decreasing end-point variability. Additionally, Christou et al. (2007) reported that young adults increased time-to-peak force variability with training without changing force-trajectory variability, and that the subjects who had the greatest improvements in time end-point accuracy had the smallest decreases in time-to-peak force variability. Poston et al. (2010) showed that trajectory variability was not associated with end-point accuracy or end-point variability for isometric goal directed contractions at 4 different force levels (20, 40, 60, and 80% of maximum force) and two different contraction speeds (time to peak of 150 ms and 1 s).

It is important to note that in all studies showed above the training that caused either adaptation and/or learning was done for periods of at most a week. Additionally, all motor task tested were highly controlled. In all studies referenced here, movements were constrained to the horizontal plane. Additionally, in most studies movements were performed within a pre-imposed time window. These facts motivate further investigation to check if the principles that are claimed in those studies are still valid when movements are trained for longer periods and are done with no space and time constraints.

To test the minimum variance theory with sports movements are valuable because in many sports an increase in the size of the neural control signal is a common training adaptation. Furthermore, probably more so than in most other sports, martial arts and/or combative sports demand its athletes to be both powerful and accurate. Thus, considering the minimum variance principle, one question arises, are martial artists and combative sports athletes able to increase movement acceleration over the years and, at the same time, either maintain or improve accuracy? As far as the authors are aware, such question has not been directly answered (Neto, 2011). Nevertheless, previous studies have shown that Kung Fu practitioners of greater experience exhibit greater hand speeds than practitioners of lesser experience (Neto, Magini, & Saba, 2007; Neto, Magini, Saba, & Pacheco, 2008). Additionally, it has been shown a significant negative linear correlation between striking force and accuracy for several strikes performed by several Kung Fu practitioners (Neto, Bolander, Pacheco, & Bir, 2009).

The first main goal of this study is to investigate the possible trade off between hand peak acceleration and accuracy and/or consistency for Kung Fu martial artists with different training experience (ranging in years) while performing maximum effort goal-directed strikes.

Force data collection of martial arts strikes is challenging since not always it is possible to have the subjects hit a force or transducer or pressure scan while data is collected (Stojsih et al., 2010). To solve this puzzle, researches may collect other physical data either from the striking hand or feet or from the object being hit. The two types of data most often collected for this purpose are acceleration using accelerometers and velocity using high speed video (Neto et. al 2007; Stojsih et al., 2010). For instance, when the surface being hit is not force sensitive, strike force may be estimated or considered correlated to the subjects hand speed before the impact and/or the target acceleration during the impact. Other variables maybe used as well, and despite the importance of understanding how these variables correlated to impact force there are not many studies on this specific subject. Furthermore, correlations of two variable across trials of the same subject or across mean values of different subjects are not necessarily similar, this will be further discussed later on in the chapter.

Thus, the second main goal of this study is to investigate the correlation among several physical variables commonly used to quantify the performance of martial arts and combative sports strikes.

Methods

Participants

Ten male and 3 female martial artists (Moy Tung Ving Tsung) volunteered to participate in the experiment. The participants had an approximate average of 4 years of martial arts training experience ranging from 1-8.5 years. The methodology was approved by the Wayne State University Human Investigation Committee, and all participants provided their informed written consent.

Goal Directed Task

After warm up, participants were requested to strike with their dominant hand 12 times a target point located in a flat surface foam pad attached to a pendulum apparatus. Participants were asked to hit as fast, as accurate and as strong as they possibly could. Before their first trial, participants were free to set their initial trunk distance to the target and feet position, this position was kept the same for all strikes. The target point was aligned with the participants' body median plane.

Data Collection

We used a load cell mounted to a steel arm that was attached to a hinge allowing it to move as a pendulum. A foam pad with a thin Acrylonitrile butadiene styrene (ABS) plastic sheet covering was placed on the striking surface of the load cell to protect the hand. The thin plastic served as a method to prevent the deformation of the foam after repeated strikes and provide a flat surface where we attached a high-speed pressure sensor matrix (model 9500, Tekscan Inc.) to determine the accuracy of the strikes. The Tekscan 9500 sensor matrix has a square sensing area of 50.41 cm^2 and contains a total of 196 sensels (3.9 sensels per cm^2). Additionally, a thin (5 cm) cushion was attached to the high-speed sensor to protect it from shear forces of the punches. A target point (1 cm^2 square) was marked into the thin cushion with its centre aligned with the geometric centre of the sensor. A Tekscan system (model 4.23I, Tekscan Inc) was used to collect the high-speed pressure data at 216 Hz. Additionally, participants hand acceleration during the strikes was obtained using three accelerometers (model 7264D, Endevco Inc.). The accelerometers were mounted onto a tri-axial block and secured to the participant's forearm before the beginning of the tests. Another tri-axial block with mounted accelerometers of the same model were attached to the load cell near its center of mass. A TDAS acquisition system (Diversified Technical Systems Inc.) was used to collect one second of acceleration data at 10,000 Hz per channel.

Data Analyses

All data were analyzed off-line using custom-written programs in Matlab® (Math Works™ Inc., Natick, Massachusetts, USA). To calculate accuracy, pressure data were used to calculate the distance in centimetres between the geometric centre of the sensor and the centre of pressure of the hand's initial contact with the striking surface. The centre of pressure of the hand's initial contact was calculated considering the pressure distribution over the sensor matrix during the first three frames of data obtained by the high speed pressure sensor matrix. Acceleration data from each accelerometer was converted to m/s^2 and low-pass filtered at 500 Hz (Butterworth order 4). Then, the magnitude of the resultant acceleration from the three orthogonal accelerations was obtained. The acceleration profile during the strike (after the beginning of movement and before the impact) were separated from the profile during the collision. The beginning of movement and the instant of the impact were determined by visual inspection in a two step process (Figure 1). First, a one second window of resultant acceleration was plotted, and by clicking twice the inspector would select a smaller window of time that comprised the acceleration during the movement and the impact (Figure 1A). Second, the smaller window would be plotted, and the inspector would determine when the strike movement began (acceleration started to increase monotonically) and when was the instant of the impact (acceleration increased rapidly) (Figure 1B). Peak acceleration were calculated considering the windows C3-C4 and C4-C2.

Variables

Several physical variables were quantified, such as: the hand peak acceleration prior to and during impact, the pendulum peak acceleration during the collision and the peak force of impact. Additionally, subjects' striking accuracy was estimated by the subject-centroid radial error (SRE), where the centroid represents a point whose coordinates are given by the average x value and average y value of the strikes performed by each subject. The SRE is given by the radial distance of the centroid from the target and is a measure of the magnitude of bias over multiple strikes from a single subject (Hancock, Butler, & Fischman, 1995). Subjects' striking consistency was estimated by the bivariate variable error (BVE). This measure is given by the square root of the 12 strikes mean squared distance from their centroid (Hancock et al., 1995). To exemplify the pressure data collected, Figure 2 shows the 12 strikes performance of two participants, one with 2 years of experience (Novice), and the other with 8.5 years of experience (Experienced).

Statistics

To achieve the first goal of the study only male subjects data were used. After the verification of the normality of the data through Kolmogorov–Smirnov tests we used Pearson's correlation to quantify the linear association among variables and between each variable and the training experience of the male participants. Hand peak acceleration was found to exhibit a lognormal distribution; thus, log values of acceleration were used. p-values less than 0.05 were considered significant. To achieve the second goal of the study data from all 13 subjects were used. Pearson's correlation were used to quantify the linear association peak force and the other variables. Analyses were performed with the SPSS 16.0 statistical package (SPSS Inc., Chicago, IL.).

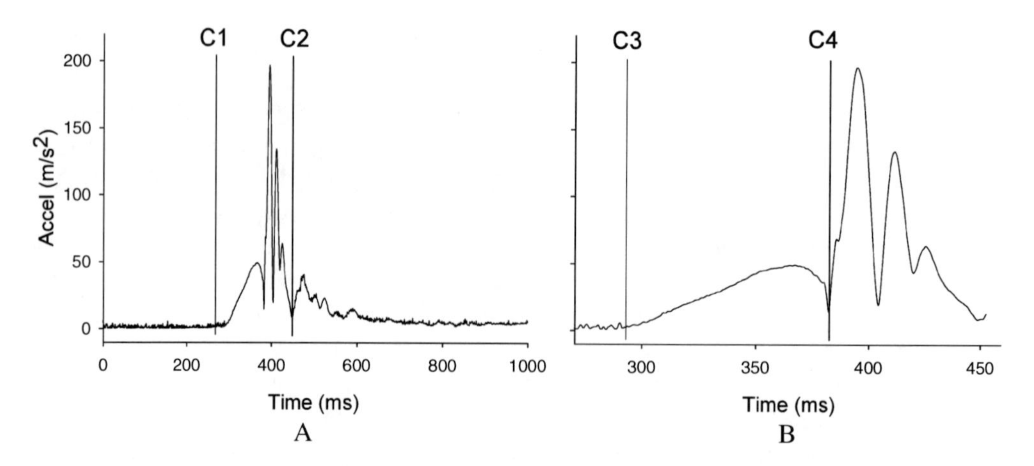

Figure 1. Example of how the beginning of the strike movements and the instant prior to impacts were determined by visual inspection. A: shows the plot of the acceleration data obtained by one participant where by clicking twice (C1 and C2) the inspector determined the next plot (B). B: with a narrower window of time around the movement and impact, the inspector determined the instant where the strike movement began (C3) and the instant prior to the impact (C4).

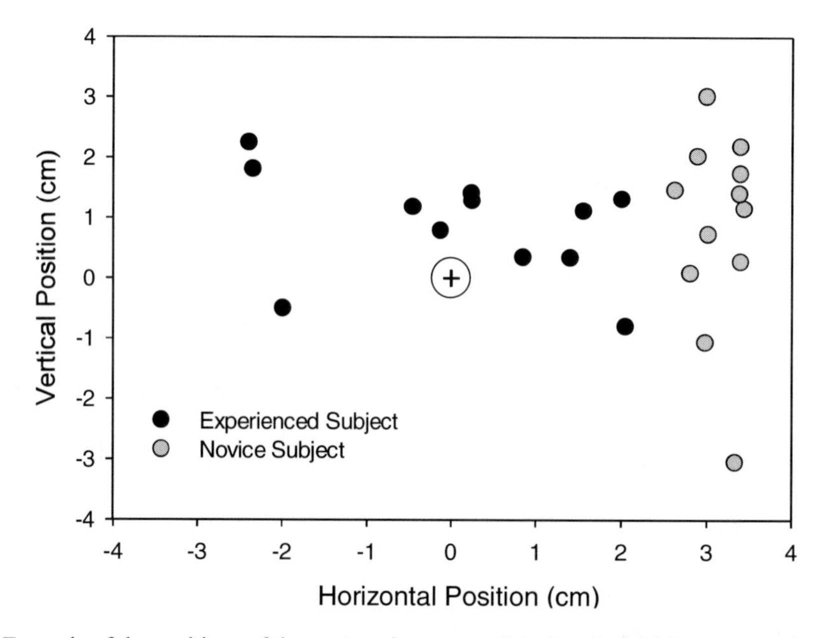

Figure 2. Example of the positions of the centre of pressure of the hand's initial contact to the target in 12 strikes of two participants, one with 2 years of experience (Novice Subject) and the other with 8.5 years of experience (Experienced Subject).

Results

The mean peak hand speed before collision was 4.11 (SD 0.61) m/s. The hand peak acceleration prior to impact was 56.2 (SD 11.9) m/s^2 and during impact was 580 (SD 116.6) m/s^2. The mean peak force of impact was 714.6 (SD 213.1) N and the peak pendulum acceleration was 75.07 (SD 28.2) m/s^2. The mean accuracy and consistency of the male participants were 2.08 (SD 0.77) cm and 1.43 (SD 0.4) cm, respectively.

For the male subjects we found a significant positive correlation between training experience and hand peak acceleration (Figure 3; r^2 = 0.655, p = 0.005). The correlation indicated that more experienced participants achieved higher peak accelerations.

Additionally, we found a significant negative correlation between training experience and accuracy (Figure 4; r^2 = 0. 621, p = 0.012). The correlation indicated that more experienced participants were also more accurate.

Finally, we found no significant correlation between training experience and consistency (Figure 5; r^2 = 0. 085, p = 0.413).

Furthermore, considering data from all 13 subjects and within-subject variations, we found that peak hand acceleration during the impact (mean r^2=0.56) followed by peak hand speed before impact (r^2=0.33) were the variables that exhibited greater correlation to peak force (Table 1). Hand peak acceleration during collision was significantly correlated to peak impact force for 10 subjects (77%), whereas hand peak speed before impact was significantly correlated to peak impact force for only 5 subjects (38%).

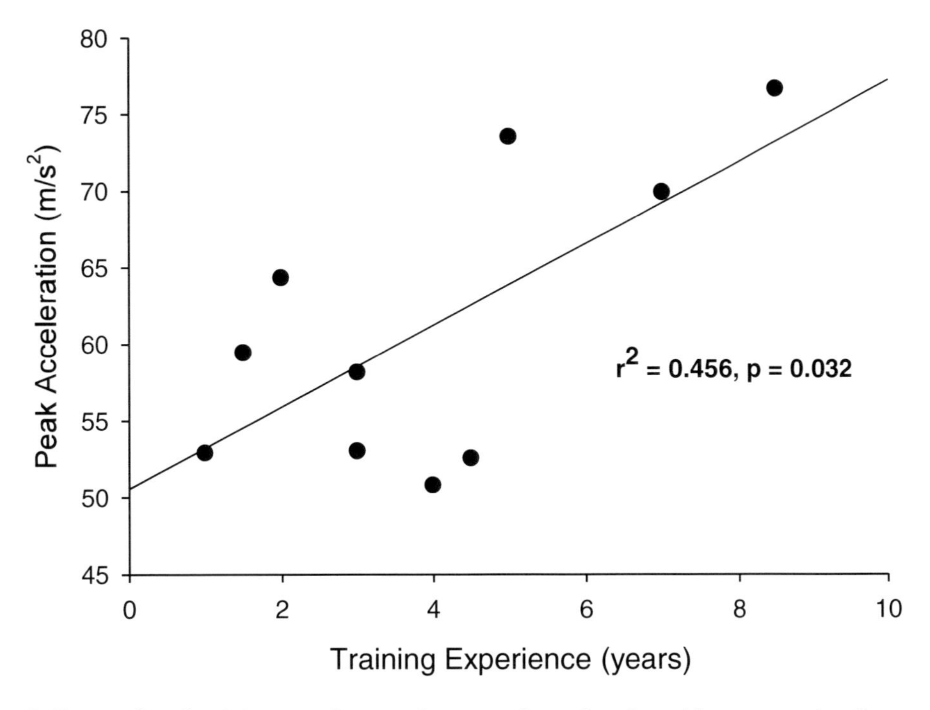

Figure 3. Scatter plot of training experience and mean peak acceleration with correspondent linear regression line showing a significant correlation between the variables (r^2 = 0.456, p = 0.032). The correlation indicates that more experienced participants exhibited higher peak accelerations.

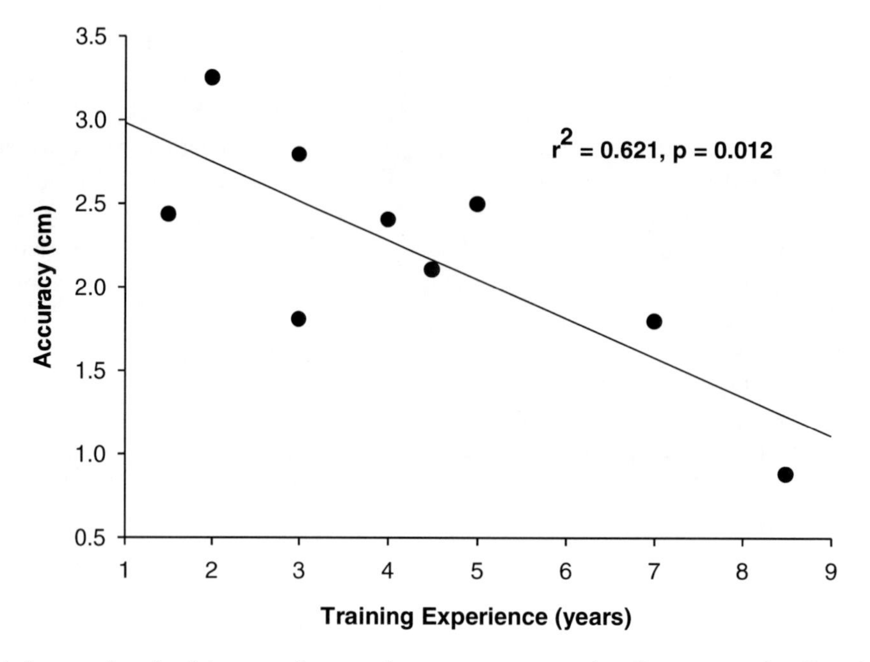

Figure 4. Scatter plot of training experience and accuracy correspondent linear regression line. A: The correlation indicates that more experienced participants were more accurate ($r^2 = 0.621$, $p = 0.012$).

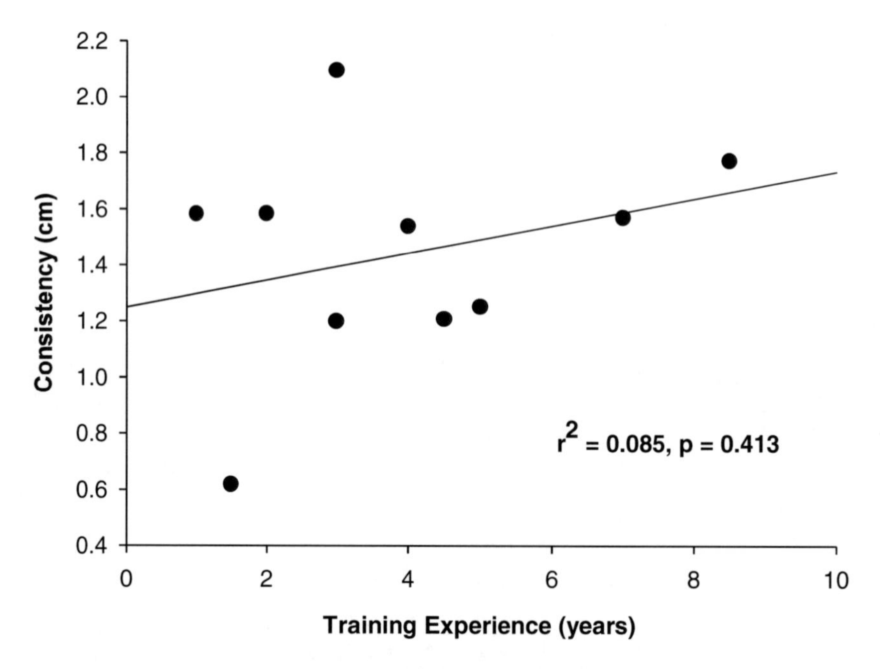

Figure 5. Scatter plot of training experience and consistency with correspondent linear regression line. The correlation indicates no significant association between training experience and consistency ($r^2 = 0.085$, $p = 0.413$).

Table 1. Correlation coefficients (r^2) and significance p-values (p) for the linear association between peak impact force and several other variables for the values obtained during 12 strikes of each subject: peak hand speed before impact (Hand Speed); peak hand acceleration before impact (Hand Acc Before); peak hand acceleration during impact (Hand Acc During); and peak pendulum acceleration during impact (Pendulum Acc)

Subjects	Hand Speed		Hand Acc Before		Hand Acc During		Pendulum Acc	
	r2	p	r2	p	r2	p	r2	p
1	0.47	0.015	0.05	0.467	0.72	0.001	0.32	0.055
2	0.08	0.378	0.75	<.001	0.16	0.191	0.01	0.811
3	0.01	0.786	0.45	0.017	0.12	0.262	0.00	0.884
4	0.25	0.118	0.04	0.544	0.58	0.006	0.31	0.075
5	0.31	0.057	0.04	0.557	0.75	<0.001	0.11	0.29
6	0.50	0.01	0.11	0.289	0.29	0.073	0.11	0.281
7	0.12	0.267	0.02	0.628	0.60	0.003	0.03	0.617
8	0.21	0.13	0.28	0.078	0.52	0.008	0.01	0.761
9	0.10	0.322	0.00	0.901	0.78	<0.001	0.02	0.637
10	0.24	0.104	0.10	0.32	0.37	0.037	0.00	0.968
11	0.54	0.006	0.20	0.142	0.77	<0.001	0.25	0.1
12	0.74	<0.001	0.01	0.706	0.79	<0.001	0.36	0.039
13	0.69	0.001	0.05	0.502	0.89	<0.001	0.60	0.003
mean	**0.33**		**0.16**		**0.56**		**0.16**	
STD	**0.24**		**0.22**		**0.25**		**0.19**	

Table 2. Correlation coefficients (r^2) and significance p-values (p) for the linear association between peak impact force and several other variables for the mean values obtained by each subject: peak hand speed before impact (Hand Speed); peak hand acceleration before impact (Hand Acc Before); peak hand acceleration during impact (Hand Acc During); and peak pendulum acceleration during impact (Pendulum Acc)

Hand Speed		Hand Acc Before		Hand Acc During		Pendulum Acc	
r2	p	r2	p	r2	p	r2	p
0.82	<0.001	0.82	<0.001	0.60	0.002	0.56	0.003

As for between-subject comparisons, we found that peak hand acceleration before impact (r^2=0.83) and again hand speed before impact (r^2=0.83) were the highest correlated measures to peak force. All variables studied exhibited a significant linear correlation to peak force. Interestingly, hand acceleration before impact, a variable that did not correlate to impact force considering within-subject trials, was highly correlated to impact force across subjects.

Discussion

The goals of this study were to investigate the possible trade off between movement peak acceleration and hand striking accuracy and consistency from martial artists with different training experience (ranging in years) and investigate the correlation between several physical variables and peak force. We found that training experience was indeed correlated to striking peak acceleration, indicating that participants with longer training experience exhibited greater acceleration prior to impact. However, greater hand acceleration did not contribute to worse performance in terms of accuracy and consistency for the more experienced participants. In fact, more experienced practitioners were more accurate and exhibited similar consistency than less experienced practitioners. Additionally, we found that the correlation between physical variables are not consistent when trials within the same subject are analysed in contrast to when mean values across subjects are analysed.

Training Experience and Consistency

Considering the minimum variance principle, we hypothesised that more experienced participants would exhibited higher peak accelerations and, in consequence, worse consistency than less experienced participants; however, our results suggest that more experienced participants presented higher peak accelerations but similar consistency to less experienced practitioners. The similar consistency across participants with different peak accelerations may reflect training induced adaptations to consistency that are happening in parallel with training induced adaptations to hand acceleration. Similar training induced adaptation has been shown for non functional tasks (Corcos, Jaric, Agarwal, & Gottlieb, 1993). Furthermore, in the task we analyzed, the entire trajectory depends on a pre-programmed strategy, and due to its short duration, no corrections are possible (Kornhuber 1971; Hallett, Shahani, & Young, 1975; Desmedt & Godaux 1978; Gordon & Ghez 1987). Thus, a possible explanation for how experienced participants are able to decrease their motor output variability is that they decrease acceleration right before contacting the target, avoiding high variability in the end of the hand trajectory (Figure 1). Such acceleration trend has been shown before for skilled Kung Fu martial artists (Neto & Magini 2008). Neto and Magini (2008) have shown that such acceleration profile is caused by a specific pattern of muscle activation and strong antagonist muscle activity prior to impact. This antagonist muscle activity before impact may be responsible to attenuate motor output variability caused by agonist muscle activation (Osu et al., 2004). This explanation has strong support also on a study done with flexion and extension movements in a visual step tracking paradigm by Darling and Cooke (1987). The authors demonstrated that forces involved in limb deceleration appeared to compensate to a greater or lesser degree for the variability in accelerative forces. Additionally, their results indicated that the linkage between accelerative and decelerative forces is progressively refined with practice resulting in decreased variability of the movement trajectory.

It is important to point out that future studies may benefit from quantifying variability aspects not only of the striking part of the movement, but also from the retracting part of the movement (after the collision). It has been suggested that motor control changes as the control modes spans a continuum from conscious to unconscious processes (Torres, 2011). In a study of martial arts strikes to imaginary targets, Torres (2011) was able to accurately predict for a randomly chosen trial whether a given technique segment was from the strike portion (more conscious) or from the retracting portion (less conscious), especially for highly trained individuals. A possible interaction between motor output variability from subjects with different expertise and movements of different levels of consciousness may be even more apparent in situations where participants actually target and hit a real point in space, as in our methodology.

Training Experience and Accuracy

We found that although more experienced participants presented higher hand peak accelerations, they were more accurate than less experienced practitioners. This result apparently contrast our previous report that there was significant negative linear correlation between striking force and accuracy for several strikes performed by several Kung Fu practitioners (Neto et al., 2009). However, differently than in the previous research, the observations done in this study relate strictly to between-subject comparisons. Our results support the idea that martial arts training may cause an improvement in striking accuracy. Similar training induced adaptation has also been shown for non functional tasks (Corcos et al., 1993). Improvements in accuracy have been associated with changes in activity within higher centers (Floyer-Lea & Matthews 2005; Hikosaka, Nakamura, Sakai, & Nakahara, 2002; Ungerleider, Doyon, & Karni, 2002). Additionally, our results supports another study of a non-striking controlled task that showed that during practice participants can increase movement speed with no detrimental effect on end-position accuracy (Darling & Cook, 1987).

Some limitations of the present study are worth discussing. First, although ideally a greater number of strikes per subject would desirable to analyze their accuracy and consistency, the striking surface was hard enough that more strikes was undesirable considering injury risk, fatigue and possible change of motor strategies during strikes due to pain or fear of getting hurt. Additionally, a larger sample size and possibly even a greater age range of training experience would certainly make our correlation finding more robust. Nevertheless, testing subjects who have consistently trained only one style of martial arts with training ages ranging evenly across almost 10 years was already challenging. Furthermore, although to ideally test the minimum variance theory and how it applies to martial arts training over several years subjects should have been tested in span of several years, such experiment is very hard to accomplish and results obtained across subjects of different training ages are worthwhile. Follow-up studies with greater number of subjects and possibly a greater number of strikes per subject are desirable.

Within-Subject and between Subject Correlations

Quantifying striking peak force is most often challenging and establishing what other physical variables should be used when trying to understand within-subject and between-subject peak force variations and no force data is available is worthwhile. In this study we quantified several other physical variables that have direct physical association to peak force. For instance, one should expect that the pendulum peak acceleration should correlate across trials to the impact force registered by the load cell, considering the accelerometers were sited near the load cell center of mass. However, our results showed that for only 2 subjects out of 13 this correlation was significant. The reason for this result is not fully known. It is possible that the pendulum acceleration was not precise enough to vary accordingly to the different forces produced trial to trial most subjects. Our results suggest that out of the variables we investigated, only hand peak acceleration during collision may be reliably used to understand impact forces variability when force data in not available. From our data hand peak acceleration was significantly correlated to peak impact force for 10 out of 13 subjects (77%). The second best variable we used was hand peak speed before impact, which was significantly correlated to peak impact force for only 5 subjects out of 13 (38%).

Different than for the correlation of variables across trials, a correlation of variables across subjects does not necessarily needs to be related to a direct physical link between the variables. It may be so that subjects with greater mean peak forces are also those with greater accuracy for example, and if that was the case the correlation across subjects of their mean peak force and mean accuracy across trials would be significant, despite the fact that these two variables are not directly linked. Understanding what variables are greatly correlated to peak force across subjects is important because results from articles on motor control, martial arts and sports performance comparing more than one group of subjects (e.g. trained vs. untrained) are based on between subjects statistics.

Our results indicate that all 4 variables investigated were significantly correlated to peak force across subjects: peak hand speed before impact, peak hand acceleration before impact, peak hand acceleration during impact, and peak pendulum acceleration during impact (Pendulum Acc). Interestingly hand acceleration before impact and hand speed before impact, that had been correlated across trials for only 2 and 5 subjects out of 13 respectively, were the variables that exhibited the greater correlation across subjects to peak force (r^2=0.83 for both variables). This result is important because hand accelerations before impact are in a much smaller range than during impact (the mean hand peak acceleration prior to impact was 56.2 m/s2 and during impact was 580 m/s^2) which makes data collection more cost efficient considering accelerometer's cost. The fact that pendulum's peak acceleration was significantly correlated to peak force across subjects support our hypothesis that the variability in peak force within most subjects' data was too small to be detected with the pendulums acceleration measures.

Conclusion

In summary, this study presents an investigation of: 1) the trade off between striking acceleration and accuracy and consistency for Kung Fu martial artists with different training experience ranging in years, 2) and the correlations among peak force and other kinematical variables. We found that greater training experience was associated to greater striking acceleration, greater accuracy, and similar consistency. Additionally, we found that hand peak acceleration during collision may be reliably used to understand impact force within-subject trial variability. Furthermore, we found that hand acceleration and speed before impact are significantly correlated to peak force for mean values across subjects.

References

Bolander, R. P., Neto, O. P. & Bir, C. A. (2009). The effects of height and distance on the force production and acceleration in martial arts strikes. *Journal Sport Science and Medicine, 8(CSSI3)*, 47-52.

Corcos, D. M., Jaric, S., Agarwal, G. C. & Gottlieb, G. L. (1993). Principles for learning single-joint movements. I- Enhanced performance by practice. *Exp Brain Res, 94*, 499-513.

Darling, W. G. & Cooke, J. D. (1987). Changes in the variability of movement trajectories with practice. *J Mot Behav, 19*(3), 291-309.

Desmedt, J. E. & Godaux, E. (1978). Ballistic skilled movements: load compensation and patterning of the motor commands. In: Desmedt, J. E. (ed) Progress in clinical neurophysiology. *Cerebral motor control in man: long loop mechanisms. Karger, Basel, 4*, 21-55.

Floyer-Lea, A. & Matthews, P. M. (2005). Distinguishable Brain Activation Networks for Short- and Long-Term Motor Skill Learning. *J Neurophysiol, 94*, 512–518.

Gordon, J. & Ghez, C. (1987). Trajectory control in targeted force impulses. II. Pulse height control. *Exp Brain Res, 67*, 241-252.

Hallett, M., Shahani, B. T. & Young, R.R. (1975). EMG analysis of stereotyped voluntary movements in man. *J Neurol Neurosurg Psychiatry, 38*, 1154-1162.

Hancock, G. R., Butler M. S. & Fischman, M. G. (1995). On the Problem of Two-Dimensional Error Scores: Measures and Analyses of Accuracy, Bias, and Consistency. *Journal of Motor Behavior, 27*(3), 241-250.

Harris, C. M. & Wolpert, D. M. (1998). Signal-dependent noise determines motor planning. *Nature, 394*, 780-784.

Hikosaka O., Nakamura K., Sakai K. & Nakahara H. (2002). Central mechanisms of motor skill learning. *Curr Opin Neurobiol, 12*, 217–222.

Kornhuber, H. H. (1971). Motor functions of cerebellum and basal ganglia: the cerebellocortical saccadic (ballistic) clock, the cerebellonuclear hold regulator, and the basal ganglia ramp (voluntary speed smooth movement) generator. *Kybernetic, 8*, 157-162.

Muller, H. & Sternad, G. (2004). Decomposition of Variability in the Execution of Goal-Oriented Tasks: Three Components of Skill Improvement. *Journal of Experimental Psychology Human Perception and Performance, 30*(1), 212–233.

Neto, O. P. (2011). *Biomechanics of Martial Arts and Combative Sports*. Hauppauge NY: Nova Science Publishers Inc.

Neto, O. P., Bolander, R., Pacheco, M. T. T. & Bir, C. A. (2009). Force, reaction time, and precision of kung fu strikes. *Perceptual and Motor Skills, 109*, 295-303.

Neto, O. P. & Magini, M. (2008). Electromiographic and kinematic characteristics of Kung Fu Yau-Man palm strike. *Journal of Electromyography and Kinesiology, 18*, 1047-1052.

Neto, O. P., Magini, M., & Saba, M. M. F. (2007). The role of effective mass and hand speed in the performance of kung fu athletes compared to non-practitioners. *Journal of Applied Biomechanics, 23*, 139-148.

Neto, O. P., Magini, M., Saba, M. M. F. & Pacheco, M. T. T. (2008). Comparison of force, power and striking efficiency for a Kung Fu strike performed by novice and experienced practitioners: Preliminary analysis. *Perceptual and Motor Skills, 1106*, 188-196.

Stojsih, S., Boitano, M., Wilhelm, M. & Bir, C. (2010). A prospective study of punch biomechanics and cognitive function for amateur boxers. *British Journal of Sports Medicine*, *44*(10), 725-30.

Osu, R., Kamimura, N., Iwasaki, H., Nakano, E., Harris, C. M., Wada, Y. & Kawato, M. (2004). Optimal impedance control for task achievement in the presence of signal-dependent noise. *J Neurophysiol*, *92*, 1199-1215.

Ungerleider, L. G. Doyon, J. & Karni, A. (2002). Imaging brain plasticity during motor skill learning. *Neurobiol Learn Mem*, *78*, 553–564.

Wolpert, D. M. (2007). Probabilistic models in human sensorimotor control. *Human Movement Science*, *26*, 511–524.

Wolpert, D. M. & Ghahramani, Z. (2000). Computational principles of movement neuroscience. *Nature Neuroscience Supplement*, *3*, 1212-1217.

In: Motor Behavior and Control: New Research
Editors: Marco Leitner and Manuel Fuchs
ISBN: 978-1-62808-142-8

Chapter 7

Assisted Cycle Therapy (ACT): Implications for Improvements in Motor Control

Shannon D. R. Ringenbach[1*], Andrew R. Albert[1], Katrin C. Lichtsinn[1], Chih-Chia (JJ) Chen[1] and Jay L. Alberts[2]
[1] Sensorimotor Development Research Lab, Program of Kinesiology, School of Nutrition and Health Promotion, Arizona State University, Tempe, AZ, US
[2] Center for Neurological Restoration, Cleveland Clinic, Cleveland, OH, US

Abstract

Assisted Cycle Therapy (ACT) is an innovative exercise in which the participant pedals on a bicycle at 35% greater than their preferred cycling rate with the assistance of a mechanical motor. Previous research in Parkinson's Disease patients found improvements in bimanual dexterity (e.g., grasping forces, interlimb coordination) and clinical measures of movement (e.g., UPDRS) after ACT but not after voluntary exercise or no exercise. Recent research with adolescents with Down syndrome found improvements in manual dexterity as measured by the Purdue Pegboard after an acute 30 minute bout of ACT but not after similar Voluntary or No exercise sessions. Improvements in the upper extremity functioning when the lower extremities were exercised suggests that changes are occurring at the cortical level to create improvements in global motor control. Possible central mechanisms include neurogenesis caused by upregulation of neurotrophic factors (e.g., BDNF) or increased sensory input to the motor cortex due to the high pedaling rate. Neurologic disorders that inhibit movement rate are suggested to benefit from ACT. The implications for improving motor, cognitive, clinical and health outcomes in several neurologic disorders will be discussed.

Keywords: Exercise, neurological disorders, executive function

---

* Corresponding Author Address: Shannon D. R. Ringenbach, Program of Kinesiology, Arizona State University, Phoenix, AZ 85004.

Assisted Cycle Therapy (ACT): Implications for Improvements in Motor Control

Voluntary exercise (VE) has long been researched as having a positive influence on mental and physical health (Penedo & Dahn, 2005). There is an emerging body of literature in healthy older adults and individuals with Alzheimer's disease indicating that exercise results in structural and functional changes in the brain (Colcombe et al., 2004; Kramer et al., 2002; Kramer et al., 2003; Kramer, Erickson, & Colcombe, 2006). These alterations in brain structure and function suggest that central nervous system function can be altered via VE in individuals with relatively normal patterns of activation within the motor cortex. However, an important limitation of Voluntary Exercise is its applicability to atypical populations, especially those who for physiological, cognitive, or behavioral reasons move too slowly to gain the aforementioned benefits.

Evolution of Forced Exercise

Forced exercise (FE) is an approach mainly used with animals in which they are exercised on a motorized treadmill at a rate greater than their voluntary exercise rate (Cotman & Berchtold, 2002; Tajiri et al., 2009; Zigmond et al., 2009). Failure to keep pace with the motorized treadmill results in a noxious stimulus (e.g., electric current). FE has been demonstrated in rodents to increase Nerve Growth Factor (NGF) (Counts & Mufson, 2005) which plays a role in promoting myelin repair (Allen & Dawbarn, 2006). Low levels of NGF are also associated with cardiovascular disease (Manni, Nikolova, Vyagova, Chaldakov, & Aloe, 2005). Thus, the clinical importance of increasing NGF is clear for improving motor control in persons with multiple sclerosis, dementia, autism, Alzhiemers, obesity, cardiovascular disease, diabetes, etc. Furthermore, research in animals shows that high intensity exercise promotes behavioral recovery in the atypical brain by modulating genes and proteins important to basal ganglia function which is crucial to voluntary motor control, cognitive, and emotional functions (Fisher et al., 2004).

The innovation of our research is to strategically and ethically apply forced exercise (FE) to humans. The first application, to our knowledge, of FE in humans was conducted in 2009 in the Alberts lab at the Cleveland Clinic (Ridgel et al., 2009). They used a tandem bicycle in which a healthy trainer maintained a pedaling rate of 80-90 RPMs, which was approximately 30% faster than the Parkinson's Disease (PD) patients' voluntary pedaling rate. Because the pedals of each rider are mechanically linked via a timing chain, the trainer and the patient were both pedaling at 80-90 RPMs. We will refer to this as Assisted Cycling (AC). Another group of PD participants pedaled at their voluntary rate. Both groups exercised for 40 minutes three times per week for eight weeks at the same relative aerobic output; the only difference in terms of the exercise performance was pedaling rate. The results of this study showed that AC improved motor function in PD patients. Specifically, following AC, clinical measures of motor control (e.g., rigidity, bradykinesia, etc.) improved, as seen by a 35% improvement in the Unified Parkinson's Disease Rating Scale (UPDRS) on the motor subscale; however no improvements in motor control were found following VE. Furthermore, the control and coordination of upper limb movements, specifically grasping forces during a bimanual dexterity task, improved following AC but not VE in PD patients. The fact that a lower extremity exercise improved upper extremity movement function indicates improvements in global motor function and suggests that improvements are happening at the cortical level.

Further research developed a stationary bicycle with a mechanical motor in which the Assisted Cycling exercise could be delivered more efficiently. The results of an eight week intervention with the stationary cycle that delivered the AC also found improvements in motor function (i.e., reduced tremor and bradykinesia) in PD patients (Ridgel, Peacock, Fickes, & Kim, 2012). Currently, a 100 patient randomized control trial is being conducted with the motorized cycle that can deliver AC three times per week for eight weeks and will use fMRI scans to measure cortical and subcortical changes that occur with improvements in motor function (Alberts, NIH-NINDS).

Possible Mechanism for Effectiveness of Assisted Cycling (AC)

Although the precise mechanism underlying improved motor and non-motor functioning following AC is unknown. An emerging hypothesis gaining support is that the increased afferent information produced by the high pedaling rate of assisted exercise paradigms produces molecular level changes at the cortical level, including up-regulation of the neurotrophic factors, including Brain Derived Neurotrophic Factor (BDNF), Nerve Growth Factor (NGF), Insulin-like Growth Factor (IGF3), Dopamine, etc. in the prefrontal and motor cortices (Cotman & Engesser-Cesar, 2002; Tajiri et al., 2009; Alberts et al., 2011). These proteins are responsible for neural plasticity, neural repair and neurogenesis, which may account for improvements in motor, cognitive and behavioral function. Since the primary role of the motor cortex is related to motor function, use-dependent forms of neuroplasticity may explain this regional specificity following an AC intervention (Petzinger et al., 2010). Recently, non-motor as well as cortical and sub-cortical changes have been shown using AC in PD patients (Alberts et al., 2011). Recent examination of the central mechanisms using fMRI procedures have shown that PD patients who performed AC produced similar changes in brain activation patterns as PD patients on medication (Beall et al., 2013). Alberts and colleagues (2011) have proposed a model of the effect of ACon the central nervous system structure and function. This may serve as a model of the treatment of other neurological conditions.

The evolution of Forced Exercise (FE), along with the development of specialized equipment has also lead to a progression of terminology with respect to this exercise paradigm. In animals it seemed appropriate to refer to this high intensity exercise as Forced Exercise (FE). For ethical reasons, in humans the term Assisted Exercise (AE) is preferred. Ridgel et al. (2012) have used the term Active-Assisted Exercise to reflect the findings that the participant must actively exert power by pushing on the pedals to receive the motor benefits of the exercise. Our lab conducts a significant amount of research on special populations and has developed the term Assisted Cycle Therapy (ACT) for this exercise paradigm, which we will use in the remainder of this chapter.

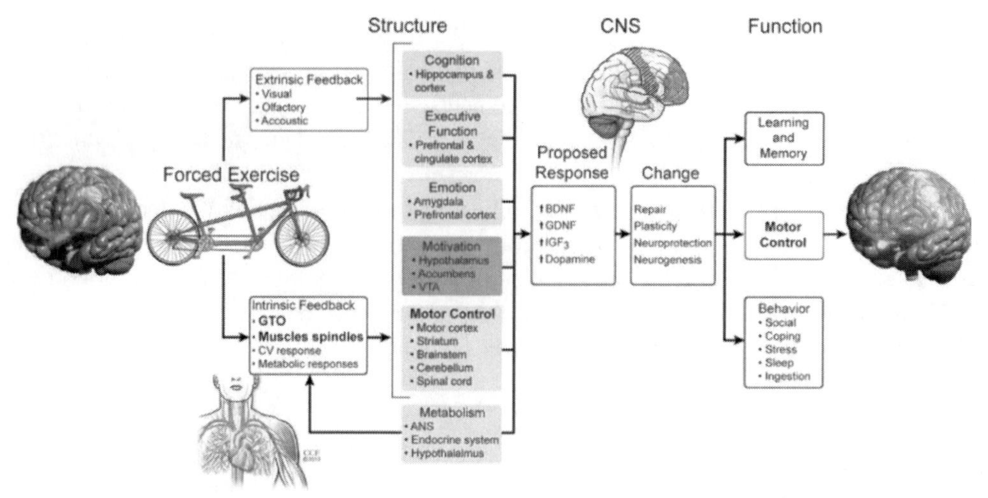

Alberts et al., 2011.

Figure 1. Schematic depicting the proposed effect of ACT on central nervous system structure and function.

There are many atypical populations with movement deficits that cannot *voluntarily* exercise at a fast enough rate to produce the neurological changes that accompany improvements in motor control. Our research seeks to translate FE data found in animals into an effective and specific therapy for special populations. Our lab has recently examined the effect of Assisted Cycle Therapy (ACT) on motor and cognitive functions in adolescence with Down syndrome (DS) and Autism Spectrum Disorder (ASD).

ACT and Down Syndrome

Current interventions for persons with DS are not achieving the desired results of improving functional tasks. Much research has examined the effect of exercise on improving physical fitness in persons with DS, however, a recent meta-analysis concluded that there was insufficient evidence to demonstrate increased physical outcomes of aerobic exercise in persons with DS (Andriolo , El Dib, Ramos, Atallah, & da Silva, 2010). The DS population has physiological (e.g., low muscle tone, congenital heart conditions, etc.), (Cioni et al., 1994; Fernhall, Pitetti, Rimmer, & McCubbin, 1996) and psychosocial factors (e.g., low motivation, sedentary lifestyle, low exercise tolerance) (Jobling & Cuskelly, 2002) that limit their exercise capabilities such that they cannot exercise at the relatively high rate of typical populations. Adolescents with DS favor sedentary activities (e.g., 90% are sedentary) (Pastore et al., 2000) which translates into 43-95% obesity (Bell & Bhate, 1992; Melville, Cooper, McGrother, Thorp & Collacott 2005; Prasher, 1995). Furthermore, approximately 61% of persons with DS have been shown to have low exercise tolerance which will affect their voluntary exercise time and intensity and thereby limit the potential motor and cognitive benefits of exercise. Our recent single session data from nine adolescents with DS confirm this. In a 30 minute VE cycling session, the average cadence for adolescents with DS was 54.6 rpm, whereas in the 30 minute ACT session, the average cadence for adolescents with

DS was 81.5 rpm. Furthermore, it was only the ACT session that resulted in pre/post improvements in *motor and cognitive* functioning (Ringenbach et al., 2012), with no changes following VE or No Exercise (NE).

Specifically, as seen in Figure 2, our results showed greater improvements in manual dexterity as assessed by the Purdue Pegboard in ACT than the VE and NE sessions which did not show any improvements in manual dexterity. Our results are consistent with research in PD patients (Alberts et al., 2009) that found manual dexterity improvements following ACT, but not VE or NE sessions. What is similar about these populations is that both PD and DS have compromised CNS functioning, which leads to deficits in movement speed, force control, and coordination.

We also found improvements in cognitive function following ACT. As can be seen in Figure 3, in adolescents with DS, cognitive planning as assessed by the Tower of London improved following ACT, but did not show improvements after VE or NE. The Tower of London paradigm has been found to activate the dorsolateral prefrontal cortex, anterior cingulate cortex, caudate nucleus, (pre)cuneus, supramarginal and angular gyrus of the parietal lobe, and frontal opercular areas of the insula (Newman, Carpenter, Varma, & Just, 2003). This finding is in keeping with proposed effect of ACT on executive function as demonstrated by activation of the prefrontal cortex in PD patients found by Alberts and colleagues (Alberts et al., 2011). In addition, simple reaction time improved following ACT, but did not show any improvements after VE or NE in adolescents with DS. The prefrontal cortex continues to be indicated as a primary area responsible for various executive function tasks, reaction time included (Koechlin & Summerfield, 2007). These results are consistent with a proposed mechanism of up-regulation of neurotrophic proteins in the prefrontal cortex following AE (Alberts et al., 2011).

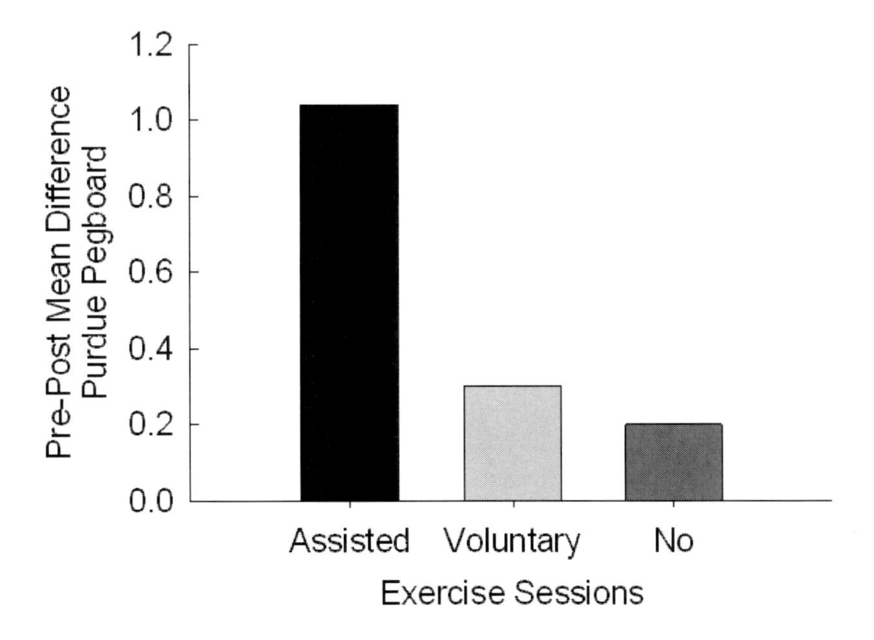

Figure 2. Pre-post difference of Purdue Pegboard as a function of exercise session in adolescents with DS.

An exciting finding for adolescents with DS is that, exercise perception improved following one ACT session and not the AE or VE sessions. We believe that during the ACT intervention, but not the VE intervention, when the participants were pedaling with the assistance of a mechanical motor, it made them feel like the exercise was less tiring, made them feel less sore, made their body feel good, and made them happier. Furthermore, because they were pedaling at such a fast rate, they perceived that they were more likely to get into shape, look better, improve their health and lose weight. Improvements in exercise perception are of particular benefit to DS populations, who typically have low motivation to exercise (Barr & Shields, 2011) and low exercise tolerance (Bell & Bhate, 1992).

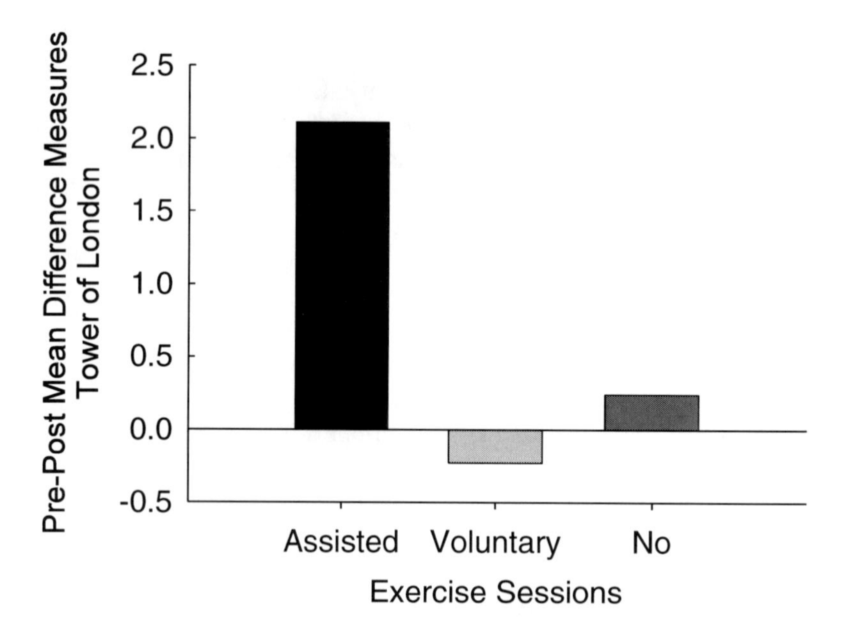

Figure 3. Pre-post difference of Tower of London as a function of exercise session in adolescents with DS.

Our results also found that after a single exercise session self-efficacy did not change. One explanation is that self-efficacy is thought to be a stable internal construct that is less influenced in short term interventions. Our previous research that found improvements in self-efficacy occurred following a cycling intervention, but not a stretching intervention, in typical participants after a seven week intervention period (Wipfli et al., 2011). Heller and colleagues found increases in performance self-efficacy following a 12 week intervention in persons with DS (Heller et al., 2004).

Similarly, we did not find improvement in digital memory span forwards (i.e., verbal short-term memory) or backwards (i.e., verbal working memory) or spatial memory span using both forwards (i.e., visual short-term memory) and backward (i.e., visual working memory) recall of Corsi Blocks. This is in keeping with recent literature that has also demonstrated the null effect of acute bouts of exercise on tests of short term memory span (Coles & Tomporowski, 2008; Ridgel, et al., 2009). However, the findings of Coles and Tomporowski (2008) demonstrated that acute bouts of exercise may facilitate the consolidation of information into long-term memory, further confounding the inconclusive evidence of the effects of exercise on different measures of memory. Similarly, the effect of chronic exercise on measures of memory seems to be widely disputed. First, no significant improvement in a memory-search task was found following both chronic aerobic and anaerobic exercise interventions (Blumenthal & Madden, 1988). However, long-term fitness has been correlated to increased relational memory, but not item memory, in preadolescent children (Chaddock, Hillman, Buck, & Cohen, 2011). Also, it has been demonstrated in rat models that following a two week period of inactivity, chronic exercise pre-disposes animals to a rapid upregulation of BDNF in the hippocampus induced by a brief exercise bout, but animals with no previous exercise do not exhibit similar upregulation of BDNF (Berchtold, Chinn, Chou, Kesslak, & Cotman, 2005). Neurotrophic proteins are proposed to induce neuroplasticity, neurogenesis, and repair, which would explain the proposed effects of BDNF on cognitive functioning (Alberts et al., 2011; Cotman & Engesser-Cesar, 2002; Tajiri et al., 2009). This finding may be beneficial to humans given the sporadic exercise habits typical of many humans, and sedentary populations like DS. Recent fMRI data has also indicated greater hippocampal activation in measures of relational memory relative to item memory (Chaddock et al., 2010) and the hippocampus is another structure that is postulated to be influenced by ACT (Alberts et al., 2011). Given the seemingly inconclusive findings of previous research, it may be that some measures of memory are more sensitive to exercise induced changes in the hippocampus than others. We are currently conducting a three times per week for eight week exercise intervention in adolescents with DS in which we can examine the effect of a long-term ACT interventions on different measures of memory, self-efficacy, and Activities of Daily Living.

ACT and Autism Spectrum Disorder (ASD)

Recently we completed a similar experiment with ten adolescents with Autism Spectrum Disorder (ASD), which is the most common neurological disorder and developmental disability affecting 1 out of every 88 children (1 in every 54 boys and 1 in every 252 girls), and is increasing in prevalence (Centers for Disease Control, 2012). ASD is characterized by impaired social interaction and communication, repetitive and stereotyped behaviors, and delays or abnormal functioning before the age of three in either social interaction, language, or symbolic or imaginative play (American Psychiatric Association, 1994). While younger children with ASD have the same physical activity levels as typical children, as children with ASD age, their physical activity levels decline. The adolescent ASD population is limited in their exercise participation due to reduced physical activity during school, social cognitive or cultural differences in typical youth sports and noninclusion in Special Olympics or Paralympics unless diagnosed with an intellectual disorder (Pann & Frey, 2006). Thus, an appropriate exercise intervention specifically for children with ASD is needed.

Because a cardinal outcome in children with ASD is motor coordination deficits, finding an appropriate exercise intervention is challenging (Fournier, Hass, Naik, Lodha, Cauraugh, 2010). However, Assisted Cycle Therapy (ACT) will eliminate clumsiness, balance, and motor coordination deficits. Exercise on a stationary bicycle reduces the balance and coordination requirements of other exercises and the solitary nature is enjoyed by this population. The experimental procedure was similar to our recent study with adolescents with

DS comparing single sessions of ACT with VE and NE separated by at least one week, except that the session length was reduced to 20 minutes of exercise to accommodate their behavioral differences. In addition, we measured a few different outcomes specific to ASD that we thought would respond to ACT based on our previous research with adolescents with DS (Ringenbach et al., 2012), and other exercise research with adolescents with ASD (Anderson-Hanley, Turreck, & Schneiderman, 2011). Enhancing motor and cognitive functioning and reducing stereotypic behavior is critical to improving activities of daily living and fostering independence and improving quality of life for persons with ASD.

As can be seen in Figure 3, inhibitory behavior as assessed by a Stroop task ($p=.034$) and to some extent set-switching ($p=.118$) and cognitive planning ($p=.071$) improved after ACT but did not change after VE or NE. Because inhibitory control and cognitive planning are both prefrontal tasks, this finding is in keeping with the proposed effect of ACT on executive function as demonstrated by activation of the prefrontal cortex in PD patients found by Alberts and colleagues (Alberts et al., 2011).

However, our results do not support the hypothesis that ACT showed greater improvements in manual dexterity, as assessed by the Purdue Pegboard than the VE and NE sessions. In fact, for adolescents with ASD, there were improvements in manual dexterity but only after the VE session and not the ACT or NE sessions. It is important to note that in both ACT and VE, the participants are exercising at the same intensity. One explanation is that their upper limb motor control did not improve following ACT is that adolescents with ASD were afraid of cycling with the motor. One of the characteristics of children with ASD is that they exhibit reduced fear in dangerous situations (e.g., snakes, running into the street) and heightened fear in harmless or new situations (e,g., ACT, interacting with other people) (American Psychiatric Association, 1994). Furthermore, Sagaspe, Schwartz and Vuilleumier (2011) suggested that the amygdala may modulate brain circuits involved in motor control by emotional signals. It can be presumed that the brain activity in the motor areas may be lower during ACT in adolescents with ASD because their movements were accompanied by a fearful emotion, relative to VE. It is hypothesized that this is caused by a dysfunction of the amygdala and its related genes in persons with ASD (Baron-Cohen et. al., 1999). As seen in Alberts and colleagues (2011) proposed model (Figure 1) ACT may improve functioning of the amygdala through increased BDNF response which may lead to improvements in social and coping behavior.

Opposite to our results with adolescents with DS, but consistent with our explanation of increased fear with ACT, exercise perception only improved following the VE session and not the ACT session in adolescents with ASD. However, adolescents with ASD do enjoy exercise as was seen with their enhanced exercise perception after the VE session. Again, we believe this stems from an increased sense of fear in uncontrollable situations such as in the ACT session due to amygdala dysfunction. Because the amygdala is one structure proposed to be influenced by ACT in Alberts et al. (2011) model we believe that over time, ACT may enhance amygdala function. Understanding exercise perception may be of particular benefit to ASD populations, who typically have poor social skills which limit their physical activity choices (Pann & Frey, 2006). Thus, future research will investigate if chronic ACT sessions can improve functioning of the amygdala and have a positive effect on social behaviors in persons with ASD.

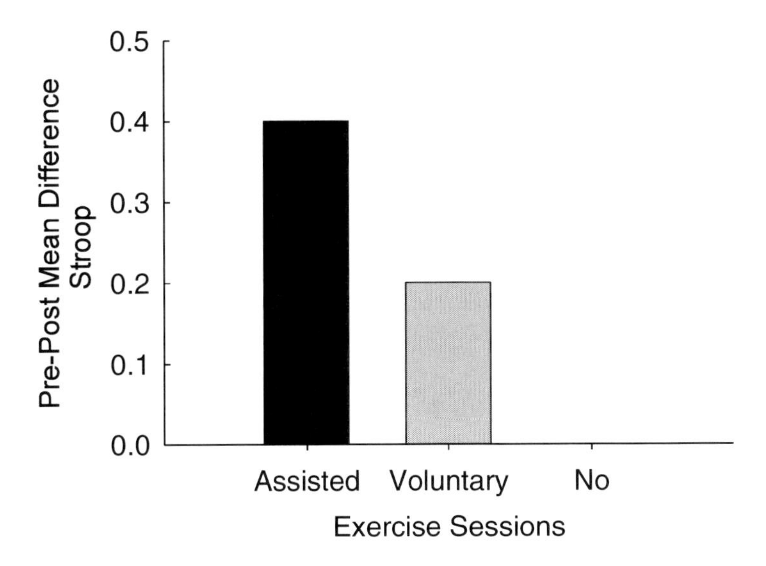

Figure 4. Pre-post difference of Stroop Pegboard as a function of exercise session in adolescents with ASD.

Conclusion

Enhancing motor and cognitive functioning is critical to improving activities of daily living and fostering independence and improving quality of life for special populations. Assisted Cycle Therapy (ACT) is a method of intervention that has had positive effects on motor and cognitive functioning in clinical populations with compromised CNS, poor motor function, low exercise motivation, and reduced cognition. Our results with several special populations reveal that positive results from this project have potential to change clinical practice, which may improve motor and cognitive functioning, as well as attitudes towards exercise for populations with neurologic disorders that inhibit movement rate. Future research will continue to examine the mechanisms responsible for improvements in global motor functioning following ACT interventions.

References

Alberts, J. L., Linder, S. M., Penko, A. L., Lowe, M. J., & Phillips, M. (2011).It is not about the bike, it is about the pedaling: forced exercise and Parkinson's disease. *Exercise and Sport Sciences Reviews, 39(4),* 177-186. doi: 10.1097/JES.0b013e31822cc71a. Review.

Allen, S. J., & Dawbarn, D. (2006). Clinical relevance of the neurotrophins and their receptors. *Clinical Science, 110 (2),* 175–191. doi:10.1042/CS20050161. PMID 16411894.

American Psychiatric Association. (1994). Pervasive developmental disorders. In Diagnostic and statistical manual of mental disorders (Fourth edition (DSM-IV)). Washington, DC:

American Psychiatric Association. 68. doi: http://www.cdc.gov/ncbddd/autism/hcp-dsm.html

Anderson-Hanley, C., Tureck, K., & Schneiderman, R. L. (2011). Autism and exergaming: effects on repetitive behaviors and cognition. *Psychology Research and Behavior Management, 4, 129-137.* doi: 10.2147/PRBM.S24016

Andriolo, R. B., El Dib, R. P., Ramos, L., Atallah, A. N., & da Silva, E. M. (2010). Aerobic exercise training programmes for improving physical and psychosocial health in adults with Down syndrome. *Cochrane Database Systematic Review, 12(5),* CD005176.

Baron-Cohen, S., Ring, H. A., Wheelwright, S., Bullmore, E. T., Brammer, M. J., Simmons, A., & Williams, S. C. (1999). Social intelligence in the normal and autistic brain: an fMRI study. *European Journal of Neuroscience, 11,* 1891-1898. doi: http://docs.autismresearchcentre.com /papers/1999_BCetal_FMRI.pdf

Barr, M. & Shields, N. (2011). Identifying the barriers and facilitators to participation in physical activity for children with Down syndrome. *Journal of Intellectual Disability Research, 55(11),* 1020-1033.

Beall, E.B., Lowe, M.J., Alberts, J. L., Frankemolle, A. M., Thota, A. K., Shah, C., & Phillips, M. D. (2012). The Effect of Forced-Exercise Therapy for Parkinson's Disease on Motor Cortex Functional Connectivity. *Archives of Physical and Medical Rehabilitatiaon, 93(11),* 2049-2054. doi: 0.1016/j.apmr.2012.05.015.

Bell, A. J., & Bhate, M. S. (1992). Prevalence of overweight and obesity in Down's syndrome and other mentally handicapped adults living in the community. *Journal of Intellect Disability Research, 36(4),* 359-364.

Berchtold, N. C., Chinn, G., Chou, M., Kesslak, J. P., & Cotman, C. W. (2005). Exercise primes a molecular memory for brain derived neurotrophic factor protein induction in the rat hippocampus. *Neuroscience, 133(3),* 853-861.

Blumenthal, J. A., & Madden, D. J. (1988). Effect of aerobic exercise training, age, and physical fitness on memory-search performance. *Psychology and Aging, 3(3),* 280-285.

Centers for Disease Control. (2012). Prevalence of Autism Spectrum Disorders—Autism and Developmental Disabilities Monitoring Network, United States, *Surveillance Summaries, 61(SS03),* 1-19.

Chaddock, L., Hillman, C. H., Buck, S. M., Cohen,N. J. (2011). Aerobic fitness and executive control of relational memory in preadolescent children. *Medicine and Science in Sport and Exercise, 43(2),* 344-349.

Cioni, M., Cocilovo, A., Di Pasquale, F., Araujo, M.B., Siqueira, C.R., & Bianco, M. Strength deficit of knee extensor muscles of individuals with Down syndrome from childhood to adolescence. *American Journal of Mental Retardation, 99(2),* 166-74.

Colcombe, S. J., Kramer, A. F., Erickson, K. I., Scalf, P., McAuley, E., Cohen, N.J., Webb, A., Jerome, G.J., & Marquez, D. X., & Elavsky, S. (2004). Cardiovascular fitness, cortical plasticity, and aging. *Proceeds of the National Academy of Science, 101,* 3316-3321.

Coles, K., & Tomporowski, P. D. (2008). Effect of acute exercise on executive processing, short-term memory, and long-term memory. *Journal of Sports Science, 26(3),* 333-344.

Cotman, C. W., & Berchtold, N. C. (2002). Exercise: A behavioral intervention to enhance brain health and plasticity. Trends in Neuroscience, 25, 295-301.

Cotman, C. W., & Engesser-Cesar, C. (2002). Exercise enhances and protects brain function. *Exercise and Sport Sciences Reviews, 30,* 75-79.

Counts, S. E., & Mufson, E. J. (2005). The role of nerve growth factor receptors in cholinergic basal forebrain degeneration in prodromal Alzheimer disease. *Journal Neuropathology and Experimental Neurology, 64(4),* 263–272.

Fernhall, B., Tymeson, G., Millar, L., & Burkett, L. (1989). Cardiovascular fitness testing and fitness levels of adolescents and adults with mental retardation including Down syndrome. *Education and Training in Mental Retardation and Developmental Disabilities, 24(2),* 133-138.

Fisher, B. E., Petzinger, G. M., Nixon, K., Hogg, E., Bremmer, S., Meshul, C.K., Jakowec, M.W. (2004). Exercise-induced behavioral recovery and neuroplasticity in the 1-methyl-4-phenyl-1,2,3,6-tetrahydropyridine- lesioned mouse basal ganglia. *Journal of Neurology Research, 77,* 378-390. doi:10.1016/j.ijcard.2004.10.041. PMID 15939120.

Fournier, K. A., Hass, C. J., Naik, S. K., Lodha, N., & Cauraugh, J. H. (2010). Motor Coordination in autism spectrum disorders: a synthesis and meta-analysis. *Journal of Autism and Developmental Disorders, 40(10),* 1227-1240. doi:http://dx.doi.org/10.1007 /s10803-010-0981-3

Heller, T., Hsieh, K., & Rimmer, J. H. (2004). Attitudinal and psychosocial outcomes of a fitness and health education program on adults with down syndrome. . *American Journal of Mental Retardation, 109(2),* 175-85.

Jobling, A., & Cuskelly, M. (2002). Life styles of adults with Down syndrome living at home. Philadelphia, PA, US: Whurr Publishers.

Koechlin, E., & Summerfield, C. (2007). An information theoretical approach to prefrontal executive function. *Trends in Cognitive Sciences, 11(8),* 229-235.

Kramer, A. F., Colcombe, S., Erickson, K., Belopolsky, A., McAuley, E., Cohen, N. J., Webb, A., Jerome, G. J., Marquez, D. X., & Wszalek, T. M. (2002). Effects of aerobic fitness training on human cortical function: A proposal. *Journal of Molecular Neuroscience,19,* 227-231.

Kramer, A. F., Colcombe, S. J., McAuley, E., Eriksen, K.I., Scalf, P., Jerome, G. J., Marquez, D. X., Elavsky, S., & Webb, A. G. (2003). Enhancing brain and cognitive function of older adults through fitness training. *Journal of Molecular Neuroscience, 20,* 213-221.

Kramer, A. F., Erickson, K. I., & Colcombe, S. J. (2006). Exercise, cognition, and the aging brain. *Journal of Applied Physiology, 101,* 1237-1242.

Manni, L., Nikolova, V., Vyagova, D., Chaldakov, G. N., & Aloe, L. (2005). Reduced plasma levels of NGF and BDNF in patients with acute coronary syndromes. *International Journal of Cardiology, 102 (1),* 169–171.Tajiri, N., Yasuhara, T.,

Melville, C. A., Cooper, S., McGrother, C. W., Thorp, C. F, & Collacott, R. (2005). Obesity in adults with Down syndrome: A case-control study. *Journal of Intellectual Disability Research, 49(2),* 125-133.

Mufson, E.J. (2005). The role of nerve growth factor receptors in cholinergic basal forebrain degeneration in prodromal Alzheimer disease. *Journal of Neuropathology and Experimental Neurology, 64 (4),* 263–272.

Newman, S. D., Carpenter, P. A., Varma, S., & Just, M. A. (2003). Frontal and parietal participation in problem solving in the Tower of London: fMRI and computational modeling of planning and high-level perception. *Neuropsychologia,41,* 1668-82.

Pann, C-Y., & Frey, G. C. (2006). Physical activity patterns in youth with autism spectrum disorders. *Journal of Autism and Developmental Disorders, 36(5),* 597-606. doi: http://dx.doi.org/10.1007/s10803-006-0101-6

Pastore, E., Marino, B., Calzolari, A., Digilio, C. M., Giannotti, A., & Turchetta, A. (2000). Clinical and cardiorespiratory assessment in children with Down syndrome without congenital heart disease. *Archives of Pediatric and Adolescent Medicine, 154(4),* 408-410.

Penedo, F. J. & Dahn, J. R. (2005). Exercise and well-being: a review of mental and physical health benefits associated with physical activity. *Current Opinions in Psychiatry, 8(2),*189-93.

Pitetti, K. H., Climstein, M., Campbell, K. D., & Barrett, P. J. (1992). The cardiovascular capacities of adults with Down syndrome: A comparative study. *Medicine and Science in Sports and Exercise, 24(1),* 13-19.

Prasher, V. P. (1995). Overweight and obesity amongst Down's syndrome adults. Journal of Intellectual Disability Research, 39(5), 437-441.

Ridgel, A. L., Peacock, C. A., Fickes, E. J., & Kim, C-H. (2012). Active-Assisted Cycling Improves Tremor and Bradykinesia in Parkinson's Disease. *Archives of Physical Medicine and Rehabilitation, 93(11),* 2049-2054. .http://dx.doi.org.ezproxy1.lib.asu.edu /10.1016/j.apmr.2012.05.015,

Ridgel , A. L., Muller, M. D., Kim, C. H., Fickes, E. J., Mera, T. O. (2013). Acute effects of passive leg cycling on upper extremity tremor and bradykinesia in Parkinson's disease. *Applied Physiology Nutrition and Metabolism, 38(2),* 194-209. doi: 10.1139/apnm-2012-0303. Epub 2013 Feb 13.

Ridgel, A. L., Vitek, J. L., & Alberts, J. L. (2009). Forced, not voluntary, exercise improves motor function in Parkinson's disease patients. *Neurorehabilitation and Neural Repair, 23(6),* 600-608.

Ringenbach, S. D. R., Chen, C-C. (JJ), Albert, A. R., Semken, K., & Semper, L. (2012). Assisted exercise improves cognitive and motor functions in persons with Down syndrome. *Journal of Sport and Exercise Psychology, 34,* S177-178.

Sagaspe, P., Schwartz, S., & Vuilleumier, P. (2011). Fear and stop: A role for the amygdale in motor inhibition by emotional signals. *Neuroimage, 55(4),* 1825-1835.

Shingo, T., Kondo, A., Yuan, W., Kadota, T., Wang, F., Baba, T., Tayra, J.T., Morimoto, T., Jing, M., Kikuchi, Y., Kuramoto, S., Agari, T, Miyoshi, Y., Fujino, H., Obata, F., Takeda, I., Furuta, T., & Date, I. () Exercise exerts neuroprotective effects on parkinson's disease model of rats. *Brain Research,1310,* 200-207.

Wipfli, B. M, Landers, D., Nagoshi, C., & Ringenbach, S. D. R. (2011). An examination of serotonin and psychological variables in the relationship between exercise and mental health. *Scandinavian Journal of Medicine and Science in Sports, 21(3),* 474-481.

Zigmond, M. J., Cameron, J. L., Leak, R. K., Mirnics, K., Russell, V. A., Smeyne, R. J., Smith, A. D. (2009). Triggering endogenous neuroprotective processes through exercise in models of dopamine deficiency. *Parkinsonism and Related Disorders, 15,* S42-45.

In: Motor Behavior and Control: New Research
Editors: Marco Leitner and Manuel Fuchs
ISBN: 978-1-62808-142-8

Chapter 8

Developing the Ability to Activate Difficult Facial Action Units Involved in Emotional Expressions

Pierre Gosselin[*]
Pierre Gosselin, School of Psychology, University of Ottawa, Ottawa, ON, Canada

Abstract

Although adults are able to activate most of the action units involved in emotional expressions voluntarily, there are some action units that have proven to be very difficult to activate. In this paper, we investigated the effect of practice on the voluntary control of three action units: cheek raiser, upper lip raiser, and lip corner depressor. Twenty young adults were given 25 training trials to activate these action units, and their performance was assessed with the Facial Action Coding System. The results indicate an effect of practice for the upper lip raiser only. As predicted, several non-target action units were activated when the participants performed the task, and they were consistent with those found in past research.

Keywords: Emotion, facial expression, motor control

Although human beings have a fairly good control of their facial musculature, this control is not perfect. Duchenne de Boulogne (1862/1990) was among the first investigators to note that some muscles, like the outer portion of *m. orbicularis oculi*, are difficult to activate voluntarily although they are always activated when people smile spontaneously. Surprisingly, the number of studies that have examined which facial action units can be activated voluntarily is quite limited so far.

Ekman, Roper and Hager (1980) examined this issue in children (between the ages of 5 and 9) and adolescents by asking them to imitate the muscular actions performed by a model

[*] Corresponding author: Pierre Gosselin, School of Psychology, University of Ottawa, 136 Jean-Jacques Lussier, Room 3002, Ottawa, Ontario, Canada, K1N 6N5. Electronic mail may be sent to pgosseli@uottawa.ca.

presented on a video monitor. They noted that nearly all of the participants were able to activate the inner brow raiser, outer brow raiser, brow lowerer, lip corner puller, and jaw drop, but very few of them could do so for the lip corner depressor and lip stretcher. Interestingly, some action units, like the upper lid raiser, lid tightener, nose wrinkler, lower lip depressor, and lip pressor were activated by most of the older children and adolescents but not by young children. Ekman et al. also found that very few participants were able to activate the inner brow raiser and outer brow raiser independently.

Recently, Gosselin, Perron and Beaupré (2010) investigated adults' voluntary control of 20 facial action units theoretically associated with happiness, fear, surprise, anger, sadness, and disgust. As in Ekman et al.'s (1980) study, the participants were instructed to imitate the muscular action performed by a model presented on a video monitor. After reading the description of the action unit and observing the corresponding video excerpts, the participants were asked to produce the movement five times consecutively while looking at themselves in a portable mirror. To optimize their performance, they were provided with feedback after each of the first four trials. The feedback consisted of suggestions as to how to improve the movement, and how to avoid non-target movements. After this practice period, they were asked to try to produce the target action unit five times in a row, in front of the mirror placed below the camera lens. Furthermore, they were encouraged to produce only the target action unit, without any other facial activity. The FACS coding of facial behavior indicated that young adults succeeded in activating 18 of the 20 target actions units, although they often coactivated other action units. For instance, few participants were able to move specifically the inner brow raiser, outer brow raiser, cheek raiser, upper lip raiser, nasolabial furrow deepener, and lip corner depressor. None of the participants were able to move specifically the inner brow raiser, and only 5% succeeded in activating the outer brow raiser and upper lip raiser without co-activating other non-target movements at the same time. The performance was slightly better for the three other action units, with 15, 25, and 30% of the participants being able to move specifically the nasolabial furrow deepener, lip corner depressor, and cheek raiser, respectively.

The findings reported by Ekman et al. (1980) and Gosselin et al. (2010) are in agreement with Duchenne de Boulogne's contention that some action units are very difficult to activate voluntarily. What we don't know is whether adults can learn to perform these difficult action units if given sufficient practice. At the present time, the available evidence concerning this question remains essentially anecdotal. Duchenne de Boulogne (1982/1990) mentions that one of the people he studied was an actor who, after much practice, was able to activate nearly all of his facial muscles. Ekman and his collaborators (Ekman, 1985) taught people how to perform difficult action units, and found that extensive practice was usually required to achieve success. However, it is not clear whether the ability to learn the difficult action units is to be found in a few talented people only or in the general population.

In the present study, we examined whether practice would allow young adults to improve their ability to activate three action units: cheek raiser, upper lip raiser, and lip corner depressor. We chose these action units because they were found to be difficult to activate in adults (Gosselin et al., 2010). Furthermore, these action units are hypothesized to be elements of displays of different basic emotions (Ekman 2003, Ekman, Friesen & Hager, 2002; Scherer & Ellgring, 2007). The cheek raiser is a component of happiness expressions, the upper lip raiser a component of disgust expressions, and the lip corner depressor a component of sadness expressions. Given the results reported by Gosselin et al., we expected a minority of participants to activate these three action units specifically in the initial stage of their training.

The second aim of our study was to examine the specificity of the voluntary control of the face. The Term specificity refers to the degree to which a person is able to activate specific muscles or groups of muscles without coactivating other unwanted muscles (Provins, 1997). Evidence concerning the specificity of the motor control of the face in children and young adults was reported by Green, Moore, Higashikawa and Steeve (2000), and in young adults only by Gosselin et al. (2010). Green et al. measured the upper lip, lower lip, and jaw movements during the production of syllables containing bilabial consonants in children and young adults. They found that young children, especially one- and two-year-olds, generally could not produce independent movements of the lower lip, upper lip, and jaw movements while trying to produce bilabial consonants. Their results also indicated significant gain in specificity during the first six years of life, with still some refinement taking place in later childhood.

Gosselin et al. (2010) found that ordinary people can voluntarily activate almost all of the 20 action units involved in emotional expressions. However, in several instances, they cannot do so without activating other non-target action units at the same time. These unwanted movements are called associated movements and are located next to the target muscular actions. The authors noted that the outer brow lowerer was an associated movement of the inner brow raiser, the inner brow raiser an associated movement of the outer brow raiser, the lip corner puller an associated movement of the cheek raiser, and the chin raiser an associated movement of the lip corner depressor.

In this study, we investigated the associated movements of the cheek raiser, upper lip raiser, and lip corner depressor. Based on Gosselin et al.'s (2010) findings, we expected the lip corner puller to be an associated movement of the cheek raiser, and the chin raiser to be an associated movement of the lip corner depressor.

Method

Participants

Twenty undergraduate and graduate students (10 men and 10 women), recruited at the University of Ottawa, participated in this study. The mean age of the sample was 23.43 years (*SD*= 6.95). All participants were French Canadian and had French as their mother tongue. The participants received $20 CAN for their participation.

Materials

The three target action units were described and illustrated on separate sheets. Each sheet explained how to perform the target action unit, and contained a picture of the appearance changes it produces. Each movement was also illustrated on video at low speed (one third of real time). The descriptions, pictures and the video excerpts were selected from the visual material of the FACS. The participants' facial behavior was recorded with a JVC super VHS videocassette recorder and camera lens. A 30 cm X 45 cm mirror was placed below the camera lens at a distance of 60 cm from the participant.

Procedure

The participants were met individually and were asked to produce three action units: cheek raiser, upper lip raiser and lip corner depressor. The action unit cheek raiser raises the cheek, causes crow's feet and wrinkles below the eye. The action unit upper lip raiser raises the upper lip and causes bend in the shape of the upper lip. Finally, the action unit lip corner depressor pulls the corners of the lips down, and produces pouching, bagging, or wrinkling of skin below the lip corners. Each action unit was produced 25 times in a row to optimize performance. Five of these productions were practice trials and 20 were test trials. Participants took between 30 and 40 minutes to complete the task. The order of production of the different action units was randomized and different for each participant. After reading the description of the action unit and observing the corresponding video excerpts, the participants were asked to produce the movement five times consecutively while looking at themselves in the mirror. An experimenter provided feedback after each of the practice trials, telling the participants how to activate the target action units, and reminding them to avoid the activation of other action units. After this practice period, they produced each target action unit 20 times in a row before the mirror. They were instructed to produce only the target action unit, without any other facial activity. The trials within each block were separated by 15 s., and the three blocks were separated from each other by two min. Feedback on performance was given during the test trials after trials 5, 8, 11, 14, and 17.

Measures

The participants' facial behavior was coded with the FACS (Ekman & Friesen, 1978; 1992) by three certified coders who were blind as to the target facial actions the participants attempted to activate. We used the FACS, as it is currently the most comprehensive and valid tool of measurement of facial activity. This system is an anatomically based measurement method that distinguishes 44 facial action units. The FACS allows one to describe any facial configuration in terms of combination of action units. Facial coding required repeated slow-motion inspection of facial behavior, and was thus time consuming. Reliability was estimated from the scoring of 25% of the material. The mean inter-rater reliability was 0.81 between coders A and B, 0.85 between coders B and C, and 0.85 between coders A and C.

Results

Percentage of Participants Succeeding in Activating the Target Action Units

Given the paucity of available evidence concerning the ability to activate facial action units, it was important to examine whether the participants' performances in the present study were similar to the performances reported by Gosselin et al. (2010). In order to do so, we looked at the percentage of participants who were able to activate the target action units at least once within the first block of five test trials. A correct score was earned if the target action unit was activated at least once during the block of test trials, without any other action unit being activated at some point during a given trial. As one can see from Figure 1, the three action units proved to be difficult to activate. Only 25%, 15%, and 20% of the participants succeeded in activating the cheek raiser, upper lip raiser, and lip corner depressor, respectively. These values are close to those reported by Gosselin et al., which were 30%, 5%, and 25%, respectively. None of the three Fisher Exact tests conducted to detect differences between the two studies were significant (all p values were >.99).

A second way of assessing participants' performances of the task was to examine their ability to activate the target action units along with other non-target action units. Performances were much better with this less demanding criterion. The rate of success for the first block of test trials was 70% for the cheek raiser, 50% for the upper lip raiser, and 60% for the lip corner depressor. Again these values were comparable to those reported by Gosselin et al. (2010), which were 60, 45, and 65%, respectively. No differences between the two studies were detected with the Fisher Exact test (all p values were >.74).

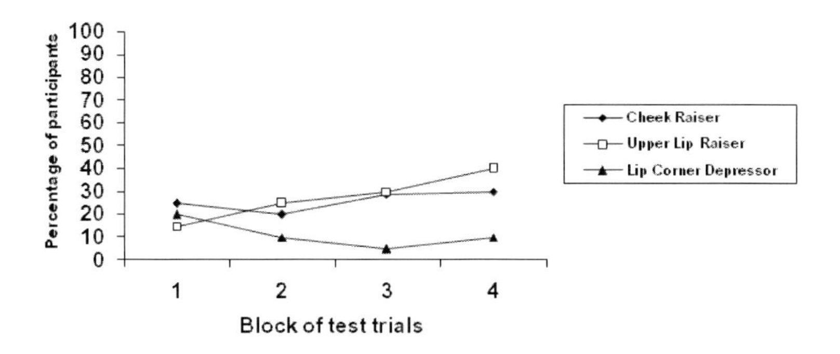

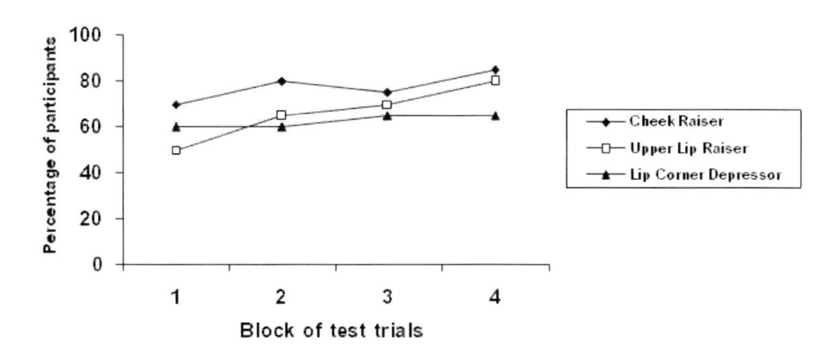

Figure 1. Percentage of participants who succeeded in activating the target action units.

Practice effect was examined by comparing the rate of success for each block of five test trials. A correct score was earned if the target action unit was activated at least once during the block of test trials, without any other action unit being activated at some point during a given trial. Although Figure 1 suggests some improvement in performance as a function of practice for the cheek raiser and upper lip raiser, the Cochran Q test showed that practice had a significant effect only on the latter case, χ^2 (3) = 9.75, $p < .02$. While only 15% of the participants initially activated the upper lip raiser, 40% did so during the fourth block of test trials. The pattern of results remained essentially the same when performance was assessed without the specificity criterion. Practice had an effect for the upper lip raiser only, χ^2 (3) = 8.33, $p < .04$. As shown in Figure 1, the percentage of participants being able to activate this action unit along with other non-target action units went from 50% (first block of test trials) to 80% (fourth block of test trials).

Performance in the Task Expressed in Terms of Percentage of Successful Trials

Another way to look at the performance in the task was to determine the percentage of successful test trials. Participants were asked to perform each target action unit 20 times. In order to assess the effect of training, we calculated the percentage of successful trials for each block of five trials. As one can see from the upper part of Figure 2, the mean performances for the specific activation of target action units were generally quite low. The highest level of success was observed for the upper lip raiser, with a mean percentage of success of 23.00 for the fourth block of trials.

Our intention was to assess the effect of training for each target action unit with a one-way repeated measure ANOVA. However, the data were not normally distributed, even after the arcsine transformation. Therefore, the data were treated with the Friedman test, which is a non-parametric test for repeated measures. The Friedman test indicated a significant effect of training for the upper lip raiser only, $\chi^2(3) = 9.62$, $p < .02$. As illustrated in Figure 2, the improvement in performance took place between the first ($M = 7.00\%$) and the second block of trials ($M = 18.00\%$).

Performance in the task was much better when it was assessed without the specificity criterion. As one can see from the bottom part of Figure 2, the mean percentages of success varied between 38.00 and 52.00 for the cheek raiser, between 33.00 and 60.00 for the upper lip raiser, and between 48.00 and 56.00 for the lip corner depressor. As the data were not normally distributed, the effect of training was assessed with a non-parametric test. The Friedman test indicated a significant effect of training for the upper lip raiser only, χ^2 (3) = 9.41, $p < .02$. Again, the improvement in performance took place between the first (M = 33.00%) and the second (M = 51.00%) block of training.

Likelihood of Associated Movements

Table 1 indicates the mean probability of occurrence of the non-target action units. Given the large number of action units distinguished in the FACS, only those with a mean probability of occurrence equal to or greater than .10 are displayed. The question we were interested in was whether these non-target action units had mean probabilities of occurrence greater than 0 (no occurrence). The data were treated with the one-sample Wilcoxon Signed-

Ranks test (directional), with the alpha error set at .006 to control for the experimentwise error rate. As shown in Table 1 (see the values of the T-statistics), five of the eight tests conducted yielded significant results. When the target action unit was the cheek raiser, two non-target action units had a mean probability of occurrence greater than 0: brow lowerer and lip corner puller. In the case of the upper lip raiser, the chin raiser and brow lowerer were the only non-target action units with a probability of occurrence greater than 0. In the case of the lip corner depressor, only the chin raiser had a probability of occurrence greater than 0.

Given there was an improvement in performance with practice for the upper lip raiser, we examined whether it was associated with a concomitant decrease in the likelihood of the chin raiser and brow lowerer. The mean likelihood of the chin raiser was .45, .43, .34, and .31, for the first, second, third and fourth block of test trials, respectively. The data were treated with the Friedman test because they were not normally distributed, nor amenable to a normal distribution. The test indicated a significant effect of practice, $\chi^2(3) = 9.47$, $p < .02$. The situation was very different for the brow lowerer, with mean likelihoods of .27, .22, .24, and .24, respectively. The Friedman test failed to yield a significant effect of training, $\chi^2(3) = 1.94$, $p < .58$.

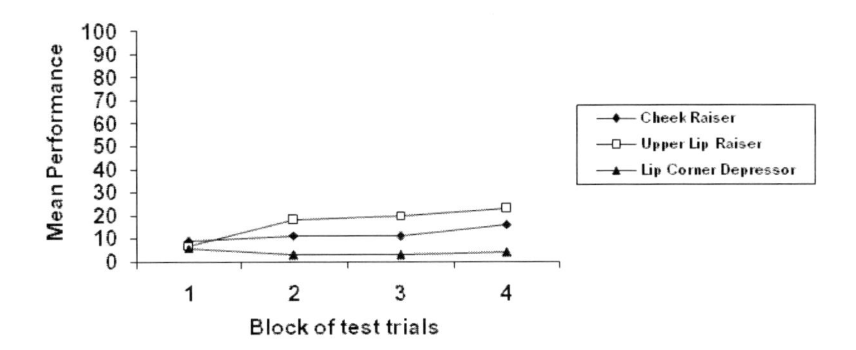

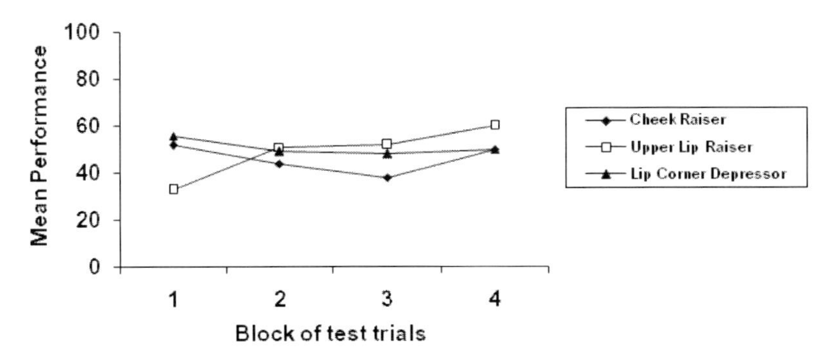

Figure 2. Mean performance (expressed in terms of percentages of successful trials) in activating the target action units.

Table 1. Likelihood of non-target action units

Target Action Unit	Non-Target Action Unit	M	SD	N	T-Statistics	p
Cheek raiser	Lip corner puller	.37	.38	18	45.50	.0002*
	Brow lowerer	.29	.34	18	27.50	.002*
Upper lip raiser	Chin raiser	.39	.44	17	33.00	.001*
	Lip corner puller	.23	.34	17	14.00	.02
	Brow lowerer	.24	.37	17	33.00	.001*
	Cheek raiser	.21	.35	17	10.50	.03
Lip corner depressor	Chin raiser	.82	.27	13	45.50	.0002*
	Upper lip raiser	.12	.24	13	3.00	.25

* Because we used the Bonferroni procedure, only the *p* values equal or less than .006 were considered significant.

Discussion

Very little is known about the extent to which it is possible to learn to activate specifically the facial action units that most people cannot perform voluntarily. We are reporting here on what we believe to be the first empirical study of the effect of training on the activation of difficult action units. First, we confirmed that the three action units we considered in this study were difficult action units to perform. Few participants were able to activate the three target action units specifically (without coactivating other action units) in the earlier stage of the training. Interestingly, the percentages of participants being able to do so were equivalent to those reported in Gosselin et al., 2010.

Second, our results indicate that only the upper lip raiser was affected by the training we provided to the participants. While 15% of the participants activated this action unit specifically at least once within the first block of trials, 40% could do so in the fourth block of trials. A similar pattern of results was found when performance was assessed without considering non-target action units. Fifty percent of the participants activated the upper lip raiser at least once during the first block of trials, and 80% could do so during the fourth block of trials. The analyses we performed on the percentages of successful trials for the effect of practice led us to the same conclusions. The training improved performance only for the upper lip raiser. The mean performance for this action unit went from 7.00 to 23.00% for the specific activation, and from 33.00 to 60.00% for the unspecific activation. Our analysis of the occurrence of non-target action units also allowed us to locate the improvement in specificity. With training, the participants were less likely to coactivate the chin raiser when they attempted to activate the upper lip raiser.

The improvement we noted for the upper lip raiser has implications for the simulation of disgust expressions. According to Ekman et al. (2002), disgust can be signaled by the nose wrinkler alone, the upper lip raiser alone, or by a combination involving the chin raiser and these two actions units. The fact that our participants were able to improve their ability to activate the upper lip raiser specifically, with a short training period, suggests that it is possible, in principle, for young adults to learn to feign disgust expressions. However, it is not clear whether they can produce such expressions in a timely manner when interacting with other people.

We did not find evidence of improvement for the cheek raiser and lip corner depressor. Our results suggest that these two action units are less amenable to voluntary control than the upper lip raiser. Given our training sessions included only 25 training trials (five practice trials plus 20 test trials), it is premature to conclude that ordinary people cannot improve their ability to activate them. Future research should focus on these two action units, and provide participants with much more training than we did.

Third, we confirmed the two predictions we made concerning the associated movements of the cheek raiser and lip corner depressor. As expected, the lip corner puller was an associated movement of the cheek raiser, and the chin raiser was an associated movement of the lip corner depressor. Furthermore, we found that the chin raiser was an associated movement of the upper lip raiser. While attempting to activate the upper lip raiser, the participants in this study had difficulty not activating the chin raiser at the same time. The mean probability of occurrence of the chin raiser was 0.39, a value close to the one (0.33) reported in Gosselin et al. (2010). It is interesting to note that the chin raiser was not found to have a probability of occurrence greater than 0 (non-occurrence) in the aforementioned study while it was in the present study. We think the discrepancy between the results of the two studies arises from the lower significance level retained by Gosselin et al. to compensate for the number of statistical tests they performed, which was nearly twice the number of tests performed in the present study.

We also found that the participants in this study were likely to coactivate the brow lowerer when they attempted to activate the cheek raiser or the upper lip raiser. These associations were not reported by Gosselin et al. (2010), which is surprising given the similarity of the two studies in terms of procedure and scoring of the facial activity. One possible explanation pertains to the number of trials we used to assess the presence of non-target movements. The measurement of the likelihood of non-target movements was based on 20 trials instead of five in Gosselin et al., which possibly gave more sensitivity to the present study.

These associated movements have different implications for the simulation of emotional expressions. The participants were likely to coactivate the lip corner puller and brow lowerer while attempting to activate the cheek raiser. We propose that the association between the lip corner puller and the cheek raiser is a facilitating factor when adults try to feign happiness, as the expression of this emotion typically involves both action units. However, the brow lowerer is not an element of spontaneous happiness expressions. We suggest that its presence could rather be a signature of feigned happiness.

We also noted that the chin raiser was an associated movement of the lip corner depressor. As the chin raiser and the lip corner depressor are both considered by emotion theorists to be elements of sadness expressions, we propose that their association is also a facilitation factor when people try to feign sadness. The same proposal can be made for the association between the chin raiser and the upper lip raiser. Given that these two action units are involved in disgust expressions, their association could be a facilitation factor when people attempt to simulate this emotion.

The recent evidence we gathered in a study with children (Gosselin, Maassarani, Younger & Perron, 2011) provided support for some of these hypotheses. School-age children were asked to portray happiness and sadness with their face as convincingly as possible, without any specific instructions concerning how to do so. The results of the study indicated that children were most likely to activate the cheek raiser and the lip corner puller when they portrayed happiness. Their portrayals of sadness included four elements: inner brow raiser, brow lowerer, lip corner depressor and chin raiser. It is interesting to note that the lip corner depressor and the chin raiser were the most common elements in the lower face.

References

Duchenne de Boulogne, G. B. (1862). *Mécanisme de la physiologie humaine ou analyse électrophysiologique de l'expression des passions.* Paris: Baillière. The mechanism of human facial expression or an electrophysiological analysis of the expression of the emotions (A. Cuthertson, trans.). New York: Cambridge University Press, 1990.

Ekman, P. (1985). *Telling lies: Clues to deceit in the marketplace,politics, and marriage.* New York: W. W. Norton.

Ekman, P. (2003). *Emotions revealed: Recognizing faces and feelings to improve communication and emotional life.* New York: Times Books.

Ekman, P. & Friesen, W. V. (1978). *Facial Action Coding System (FACS): A technique for the measurement of facial action.* Palo Alto: Consulting Psychologists Press.

Ekman, P. & Friesen, W. V. (1992). FACS update document. Available from Paul Ekman: http//: www.paulekman.com.

Ekman, P., Friesen, W. V. & Hager, J. C. (2002). *The Facial Action Coding System CD-ROM.* Salt Lake City, UT: Research Nexus.

Ekman, P., Roper, G. & Hager, J. C. (1980). Deliberate facial movement. *Child Development, 51*, 886-891.

Gosselin, P., Perron, M. & Beaupré, M. (2010). The voluntary control of facial action units in adults. *Emotion, 10*, 266-271.

Gosselin, P., Massarani, R., Younger, A. & Perron, M. (2011). Children's deliberate control of facial action units involved in sad and happy expressions. *Journal of Nonverbal Behavior, 35*, 225-242.

Green, J. R., Moore, C. A., Higashikawa, M. & Steeve, R. W. (2000). The physiologic development of speech motor control: Lip and jaw coordination. *Journal of Speech, Language, and Hearing Research, 43*, 239-255.

Provins, K. A. (1997). The specificity of motor skill and manual asymmetry: A review of the evidence and its implications. *Journal of Motor Behavior, 29*, 183-192.

Scherer, K, R. & Ellgring, H. (2007). Are facial expressions of emotion produced by categorical affect programs or dynamically driven by appraisal? *Emotion, 7,* 113-130.

In: Motor Behavior and Control: New Research
Editors: Marco Leitner and Manuel Fuchs
ISBN: 978-1-62808-142-8

Chapter 9

Coordination Thermodynamics: Theory and Applications

T. D. Frank **[*] *and D. G. Dotov***
Center for the Ecological Study of Perception and Action
Department of Psychology
University of Connecticut, Storrs, US

Abstract

Human motor behavior frequently requires a large degree of coordination. Catching requires motor coordination with an environmental stimulus. Walking and stair climbing requires inter-limb coordination. Motor activities in groups of people such as team sports activities require between-persons coordination of movements. The deterministic laws of human motor coordination have been successfully described in terms of dynamical systems. In contrast, the formulation of thermodynamic laws of motor coordination is still a challenge for modern day science. The reason for this is that a general thermodynamic theory for non-equilibrium systems such as human motor control systems is not available. In order to address this problem, it has recently been suggested to consider stochastic descriptions of human motor coordination that are grounded in linear non-equilibriumthermodynamics. Two special cases are the socalled canonical-dissipative approach to uni-manual motor control problems (Frank, Dotov, Turvey, 2010; Dotov and Frank, 2011) and the mean field theoretical approach to motor coordination in groups (Frank, Richardson, 2010; Richardson et al., 2012).

The general theoretical framework of a theory of coordination thermodynamics is outlined. In addition, the coordination thermodynamical perspective is illustrated for two experimental studies on single person motor coordination and between-persons motor coordination of groups.

Keywords: Motor coordination; uni-manual movements; synchronization; thermodynamics

[*] E-mail address: till.frank@uconn.edu.

1. Introduction

Motor coordination is an essential part of human motor control. For example, motor coordination is required during stair climbing. In general, the pattern may include aspects of the environment. Reaching towards a moving target and catching requires coordination of body movements with an external stimulus. Passing an object from one person to another, feeding an infant with a spoon, or pair dancing require the coordination of movements between two people. Team sports activities such as soccer and football involve the coordinated movements of groups of people.

Performance of a coordinated motor activity involves the activity of a plenitude of units (e.g., muscles, receptors, and neurons) on different spatial and temporal scales. In general, the muscular and skeletal system as well as the nervous system is involved. How can these units be activated to bring about a coordinated movement? It has been suggested that the units are bound together to synergies [5]. A synergy may be understood as a composition of individual units that are assembled to a larger unit, where there is both feed-forward control from the individual units to the larger unit and feedback control from the larger unit to the individual units. The individual units affect the behavior of the larger unit and the larger unit affects the behavior of the individual units. The synergy is due to circular causality. Such self-regulated organizations or synergies are known to exist as dissipative structures. Dissipative structures are spatio-temporal structures in systems that operate far from thermal equilibrium and are typically pumped by energy or an inflow of matter. Therefore it has been argued that the underlying basis of motor coordination is the formation of dissipative structures [42]. Dissipative structures in turn are instances of self-organization. In view of this observation, it does not come as a surprise that several research groups have advocated that the binding of the plenitude of muscular, skeletal, and nervous units to a whole organization that exhibits a coordinated motor activity is a self-organization process [2, 33, 39, 65]. In this context, dynamical systems theory is a promising mathematical tool that can capture key aspects of the self-organization of motor coordination both for coordinated single limb and multi-limb movements of a single actor (see e.g. [33, 39, 36]) and coordinated motor activity between people (see e.g. [58, 60]).

As mentioned above, coordination has been placed into the context of dissipative structures. Dissipative indicates that there is an inflow and outflow of energy and matter such that disorder is created in order to build-up and maintain the structure under consideration. Technically speaking, entropy – as a measure for disorder – is produced. In doing so, a connection between coordination and thermodynamics has been pointed out [42]. Similar to this proposal, it has been pointed out that a thermodynamic perspective can be developed on the basis of a stochastic description of motor behavior [24]. This argument will be developed in the subsequent sections. Figure 1A outlines the approach. First of all, statistical mechanics provides a statistical interpretation of thermodynamical variables such as energy and entropy [35, 41, 53, 67]. In addition, statistical mechanics allows for mathematically rigorous derivations of relations that hold between thermodynamical variables. Second, transport theory based for example on the Boltzmann equation includes to a certain extent statistical mechanics as a special case. Moreover, transport theory provides a theoretical framework to derive under certain circumstances the second law of thermodynamics (the law about the increase of entropy in isolated systems) from a dynamical systems perspective (see e.g., the H-theorem in Ref. [35, 53, 67]). Similarly, transport theory based on diffusion equations in combination with linear non-equilibrium thermodynamics allows for a statistical mechanics approach to thermodynamical quantities and accounts for the second law of thermodynamics [9, 18]. Finally, the theory of stochastic processes includes transport theory as a special case, on the one hand, and, on the other hand, can be regarded as a stochastic generalization of dynamical systems theory and therefore provides a mathematical framework for describing motor coordination and motor behavior. As a conclusion, a thermodynamical perspective of motor behavior in general and motor coordination in particular can be pursued by exploiting the concepts of the theory of stochastic processes, linear non-equilibrium thermodynamics, and statistical mechanics, see Figure 1B.

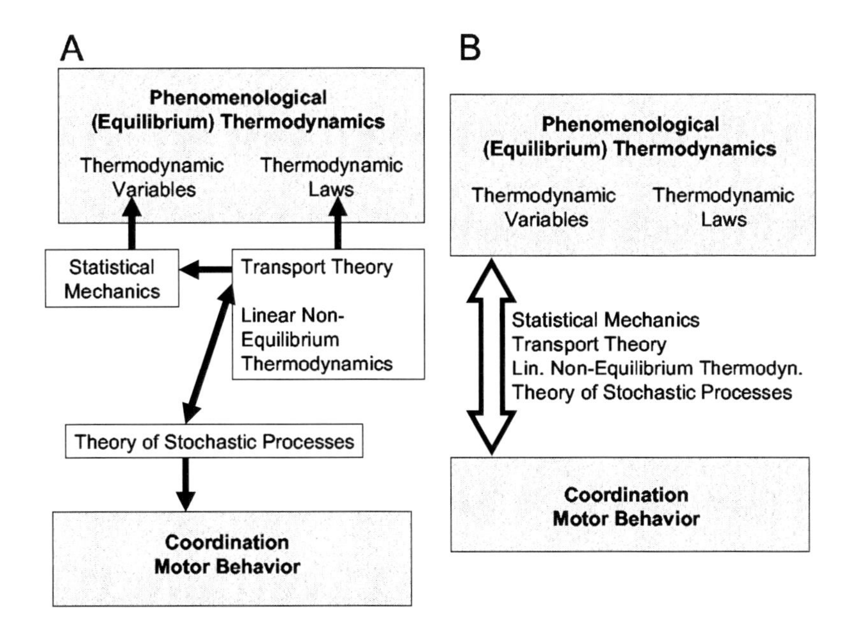

Figure 1. Outline of a thermodynamical approach to human motor coordination and motor behavior. See text for details.

Once a link between stochastic motor coordination and thermodynamics has been established, then a particularly promising thermodynamical concept is the notion of an energy form called 'free energy'. According to the Landau theory of phase transitions [45], an equilibrium system exhibiting a phase transition towards an ordered state settles down in a free energy minimum [52, 62]. Figure 2A illustrates this issue for phase transitions in nematic liquid crystals. Nematic liquid crystals exhibit two phases: the isotropic phase with microscopic disorder and the nematic phase with some degree of microscopic order [8, 62]. The amount of order can be quantified by the so-called Maier-Saupe order parameter S_{MS} (where the subindex MS stands for Maier-Saupe). For relatively high temperatures T, the free energy exhibits a minimum at $S_{\mathrm{MS}} = 0$. The liquid crystal does not exhibit microscopic order and is in its isotropic phase. At the critical temperature $T(crit)$ the minimum at $S_{\rm MS} = 0$ becomes a saddle-point and the free energy exhibits a global minimum at $S_{\rm SM} > 0$. The liquid crystal at $S_{\rm SM} > 0$ exhibits some degree of microscopic order and is in its nematic phase. In general, equilibrium systems are characterized by free energy functions that depend on control parameters such as the temperature. The free energy functions exhibit minima at certain values. The locations of the free energy minima determine the phases or states of the systems. In addition to the principle of free energy minimization, the Landau theory yields another interesting result. As illustrated in Fig. 2B, the minimal free energy decays when the control parameter (here the temperature) is scaled further and further away from the phase transition point.

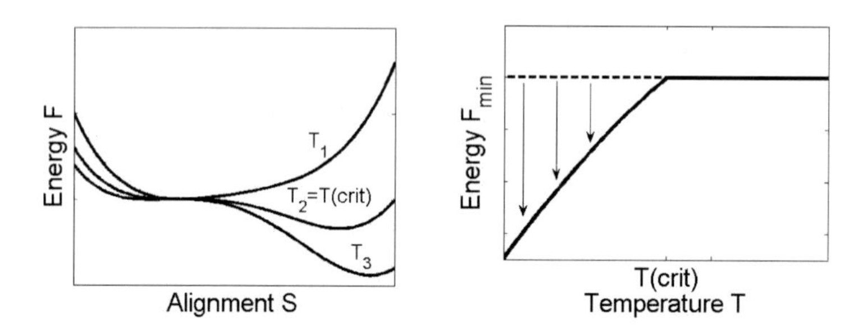

Figure 2. Free energy F in the Landau theory illustrated by means of the classical order parameter model of nematic liquid crystals. Left: free energy minimization principle. Right: free energy decays when the control parameter (i.e., here the temperature) is scaled away from the critical value. $F(S_{\rm MS})$ was calculated from $F(z) = F_0 + (T - T(crit))z^2 + cz^3 + dz^4$ with $F_0 = 0$, $c = -2$, and $d = 1.0$, see also Ref. [62]. Temperatures: $T_1 - T(crit) = 2$, $T_2 - T(crit) = 0$, $T_3 - T(crit) = -1$. Illustration in arbitrary units.

As we will show below, the thermodynamical approach to motor coordination as outlined in Figure 1 is consistent with these two thermodynamical aspects of the Landau theory: the free energy minimization and the decay of the free energy minimum as a function of the control parameter.

2. Stochastic Coordination and Motor Behavior: Thermodynamic Perspective

In this section, a theoretical framework is presented that allows to discuss stochastic motor control processes from a thermodynamics perspective.

2.1. Theory of Stochastic Processes

Let $\mathbf{x} = (x_1, \ldots, x_N)$ denote a N-dimensional state vector describing the state of a motor control system. The state is assumed to evolve in time t such that $\mathbf{x}(t)$ describes the motor behavior of an human actor as a function of time. The motor pattern is assumed to exhibit some degree of erraticness. That is, taking motor control variability into account, the function $\mathbf{x}(t)$ becomes a stochastic process. A general class of time-continuous stochastic processes is the class of Markov diffusion processes [28, 57]. We assume that $\mathbf{x}$ belongs to this class of processes. Accordingly, $\mathbf{x}(t)$ satisfies a Langevin equation [28, 57]

$$\frac{\mathrm{d}}{\mathrm{d}t}\mathbf{x}_r = \underbrace{\mathbf{h}}_{\text{deterministic}} + \underbrace{g \cdot \mathbf{\Gamma}_r(t)}_{\text{random}} . \tag{1}$$

The vector-valued function $\mathbf{h} = (h_1, \ldots, h_N)$ is a deterministic force and describes the change of the state $\mathbf{x}$ due to deterministic influences. The second term on the right-hand side constitutes a fluctuating force and accounts for random impacts inducing a change of $\mathbf{x}$. The fluctuating force is a product of two components: the Langevin force $\mathbf{\Gamma}_r = (\Gamma_{1,r}, \ldots, \Gamma_{N,r})$, which is a normalized fluctuating force [28, 57] and g which is a $N \times N$ weight matrix. The sub-index r indicates that we are dealing with the r-th (hypothetical) repetition of the process under consideration. We call $\mathbf{x}_r(t)$ the r-th realization of the process. Expectation values are calculated by averaging with respect to an infinitely large ensemble of realizations. For example, the mean value can be computed from $\langle \mathbf{x} \rangle_r = N^{-1} \sum_{r=1}^{N} \mathbf{x}_r$ in the limiting case $N \rightarrow \infty$. Here, the operator $\langle \cdot \rangle_r$ denotes ensemble averaging. The realizations of the process are distributed at every time point t according to a probability density $W(\mathbf{x}, t)$. The probability density can formally be computed from the ensemble average of the Dirac delta function $\delta(\cdot)$ like

$$W(\mathbf{x}, t) = \langle \delta(\mathbf{x} - \mathbf{x}_r(t)) \rangle_r . \tag{2}$$

2.2. Thermodynamic Variables

The aim is to relate the stochastic motor control dynamics to thermodynamic variables. To this end, thermodynamic variables used in the thermodynamics of equilibrium systems will be adopted to capture thermodynamical aspects of non-equilibrium motor control systems. Table 1 summarizes the key steps. In order to establish a link between thermodynamics and stochastic processes describing motor patterns, we introduce the notions of energy and entropy.

First, we assume that the dynamical description of the motor behavior by means of a vector $\mathbf{x}$ allows us to define a function that can be interpreted as an energy function.

For example, let us assume with study an oscillatory movement of a single limb that can approximately be described by an harmonic oscillator equation

$$\frac{\mathrm{d}^2}{\mathrm{d}t^2} x_r = -k\, x_r \tag{3}$$

with $k > 0$. This evolution equation involves the so-called Hamiltonian energy function $H(\mathbf{x}_r) = v_r^2/2 + kx_r^2/2$ with $\mathbf{x} = (x_1, x_2) = (x, v)$ and can be written like $\mathrm{d}x_r/\mathrm{d}t = v_r = \partial H/\partial v_r$, and $\mathrm{d}v_r/\mathrm{d}t = -kx_r = -\partial H/\partial x_r$. Eq. (3) may be generalized to account for random effects. In particular, the generalized version may assume the form of the Langevin equation (1). In the stochastic case, we can introduce a second energy measure: the averaged energy U defined by $U = \int_x \int_v H(x, v)W(x, v)\mathrm{d}x\mathrm{d}v$, where $W(x, v) = W(\mathbf{x})$ is the probability density in the two-dimensional state space.

In general, we assume that for the motor control problem at hand a function E can be defined that can be interpreted as an energy function. The energy function depends on the state $\mathbf{x}$ such that it represents a map of $\mathbf{x}$ to a real number:

$$E \, : \, \mathbf{x} \rightarrow E(\mathbf{x}, \cdot) \in \mathbb{R} \, . \tag{4}$$

As indicated by $(\cdot)$ the energy function may depend on a set of (fixed, time-dependent, or self-generated) parameters as well. Next, the averaged energy U is introduced by

$$U = \int E(\mathbf{x}, \cdot) W(\mathbf{x}) \, \mathrm{d}^N x \, . \tag{5}$$

The entropy S is defined by

$$S = -\int W \ln W \, \mathrm{d}^N x \, . \tag{6}$$

The definitions of U and S by Eqs. (5) and (6) are derived from statistical mechanics [35, 67]. However, there is a slightly different interpretation. In statistical mechanics the probability f represent the probability density of an infinitely large ensemble of identical systems. That is, U and S as defined by Eqs. (5) and (6) are considered a measures of many-particle systems. In contrast, we consider an infinitely large ensemble of realizations of a given motor pattern. That is, the hypothetical case is considered in which a particular experiment is repeated again and again and in doing so an arbitrary large data set is generated.

In this context, note that the entropy S is a measure for variability and 'disorder'. If the realizations $r = 1, \ldots, R$ of the motor pattern under consideration look similar (low variability, low 'disorder'), then S assumes small numbers or even negative numbers. In contrast, if the motor patterns obtained by repeating the same experiment do not look similar (high variability, high 'disorder'), then the entropy S assumes a large, positive number. In particular, let us consider the ensemble of realizations $\mathbf{x}_r$ under two different conditions A and B such that the realizations are distributed like W_A and W_B, respectively. If $S(A) > S(B)$ then this indicates that the 'disorder' or variability as measured by S is higher under condition A than under condition B.

Table 1. Thermodynamical variables capturing the impacts of different types of stochastic processes underlying motor coordination.

Subprocess Level 1	Conservative	Non-Conservative	
Subprocess Level 2		External	Internal
Thermodynamic variables	Process energy U	Process energy U Process entropy S	Process energy F Process entropy S
Impact	Invariant $\mathrm{d}U = 0$	Change $\mathrm{d}U$, $\mathrm{d}_{\text{ext}}S$	Uni-directional change $\mathrm{d}F \leq 0$, $\mathrm{d}_{\text{int}}S \geq 0$
Motor behavior	Asymptotic dynamics	Transient dynamics & Stability	Erraticness
		Variability	

2.3. Linear Non-Equilibrium Thermodynamics

How can we connect the stochastic description via Langevin equations (1) with the thermodynamical variables U and S defined by Eqs. (5) and (6)? To begin with, we assume that we can distinguish between two kinds of components that we will refer to as conservative and non-conservative. The conservative component or subprocess is the component determined by a conservative force **I**. Accordingly, the conservative subprocess in isolation satisfies the deterministic evolution equation

$$\frac{\mathrm{d}}{\mathrm{d}t}\mathbf{x}_r = \mathbf{I}(\mathbf{x}_r) \tag{7}$$

The conservative force, by definition, is orthogonal to the gradient of the energy function E, that is, we have $\nabla E \cdot \mathbf{I} = 0$ (here $\nabla = (\partial/\partial x_1, \ldots, \partial/\partial x_N)$. This implies that the energy is invariant under the dynamics defined by Eq. (7): $\mathrm{d}E/\mathrm{d}t = 0$. In addition, the force is divergence free: $\nabla \cdot \mathbf{I} = 0$. That is, there are neither sources nor sinks. An important implication of this requirement is that the probability density W satisfies the continuity equation

$$\frac{\partial}{\partial t}W = -\nabla \cdot \mathbf{J}_I \tag{8}$$

involving the flux or probability current

$$\mathbf{J}_I = \mathbf{I}\,W\,. \tag{9}$$

Let us turn next to the non-conservative component of the stochastic process. This component is composed of subprocesses that may induce a change of the process energies E and U.

Energy variations ΔU can be addressed together with entropy variations ΔS. To this end, we apply the concepts of linear non-equilibrium thermodynamics [10, 18, 29, 40] to examine the nature of changes ΔS. Let $d_{\text{ext}}S$ and $d_{\text{int}}S$ denote the changes of S due to external and internal processes, respectively. Then the total change of S is given by [10, 18, 29, 40]

$$\mathrm{d}S = \mathrm{d}_{\text{ext}}S + \mathrm{d}_{\text{int}}S\,. \tag{10}$$

For systems close to thermodynamical equilibrium for which processes do not induce a temperature change (i.e., isothermal processes), reversible changes affect the energy U like

$\mathrm{d}U = T\mathrm{d}_{\mathrm{rev}}S$. We assume that the same kind of relationship holds for external stochastic subprocesses: $\mathrm{d}U = T\mathrm{d}_{\mathrm{ext}}S$. In the context of stochastic processes, however, it is useful to replace the temperature measure T by an alternative measures Q [18]. The variable $Q \geq 0$ measures the overall strength of the fluctuating force affecting the stochastic process under consideration. This is consistent with the gas-kinetical interpretation of the temperature T that states that T is a measure for the kinetical energy of gas particles such that at higher temperatures the particles exhibit a 'more volatile' random motion than at lower temperatures. We put

$$\mathrm{d}U = Q\mathrm{d}_{\mathrm{ext}}S\,. \tag{11}$$

At this stage it is useful to introduce a second energy measure F. While U is related to entropy changes induced by external processes, see Eq. (11), the energy measure F is defined such that it is related to entropy changes induced by internal processes:

$$\mathrm{d}F = -Q\mathrm{d}_{\mathrm{int}}S\,. \tag{12}$$

Note that in Eq. (12) there is a minus sign which allows us to relate F to a thermodynamic variable used in phenomenological thermodynamics (see below). As a consequence of the second law of thermodynamics, internal processes can only results in entropy increase, which implies $\mathrm{d}_{\mathrm{int}}S \geq 0$ and $\mathrm{d}F \leq 0$, as indicated in Table 1. Substituting Eqs. (11) and (12) into Eq. (10), we obtain $Q\mathrm{d}S = \mathrm{d}U - \mathrm{d}F$. Taking into account that Q is a fixed parameter in our context, integration yields

$$F = U - Q\,S + C \;\rightarrow\; F = U - Q\,S \tag{13}$$

up to an integration constant C that is neglected as indicated above (i.e., we put $C = 0$). In the case of systems close to thermodynamic equilibrium (i.e., if we put $Q = T$), energy measures U and F satisfying Eq. (13) are known as 'internal energy' and 'free energy' (or Helmholtz free energy), respectively [35, 67]. We will make use of these notations also for the non-equilibrium case described by Eq. (13) in which U, S, F are thermodynamical variables of a stochastic process of the form (1) describing human motor coordination.

Let us proceed to close the link between Eq. (1) and the thermodynamical variables U, S, and F. To this end, we continue to work within the framework of linear non-equilibrium thermodynamics. Linear non-equilibrium thermodynamics assumes that processes exhibit thermodynamic forces and thermodynamical fluxes. Let $\mathbf{X}_{\mathrm{th}} = (X_1, \ldots, X_N)$ and $\mathbf{J}_{\mathrm{th}} = (J_1, \ldots, J_N)$ denote the vector-value thermodynamic force and flux of a process. The components X_K and J_k point along the coordinates x_k of the state space of the process. Let r denote the rate of entropy production due to irreversible changes. Then for close to equilibrium systems r is proportional to the product of force and flux and the proportionality factor is the reciprocal temperature (i.e., $1/T$). Likewise, we put

$$r(\mathbf{x}) = \frac{1}{Q}\mathbf{X}_{\mathrm{th}}(\mathbf{x}) \cdot \mathbf{J}_{\mathrm{th}}(\mathbf{x})\,. \tag{14}$$

As indicated, $r(\mathbf{x})$ is a local measure of the rate of irreversible entropy change. In order to determine $\mathrm{d}_{\mathrm{irr}}S/\mathrm{d}t$ we need to integrate over the whole state space. Thus, we obtain

$$\frac{\mathrm{d}_{\mathrm{irr}}S}{\mathrm{d}t} = \int r(\mathbf{x})\,\mathrm{d}^N x = \frac{1}{Q}\int \mathbf{X}_{\mathrm{th}}(\mathbf{x}) \cdot \mathbf{J}_{\mathrm{th}}(\mathbf{x})\,\mathrm{d}^N x\,. \tag{15}$$

We identify this change of entropy due to irreversible processes with the change of entropy due to internal processes: $\mathrm{d}_{\mathrm{irr}} S = \mathrm{d}_{\mathrm{int}} S$. Substituting Eq. (12) into Eq. (15), we obtain

$$\frac{\mathrm{d}}{\mathrm{d}t} F = -\int \mathbf{X}_{\mathrm{th}}(\mathbf{x}) \cdot \mathbf{J}_{\mathrm{th}}(\mathbf{x})\, \mathrm{d}^N x \, . \tag{16}$$

2.4. Connection Between Theory of Stochastic Processes and Linear Non-Equilibrium Thermodynamics

Above, we have shown that the deterministic, conservative process (7) satisfies the continuity equation (8). Likewise, we assume that the stochastic process (1) satisfies a continuity equation involving the probability density W on the one hand and the thermodynamic flux $\mathbf{J}_{\mathrm{th}}$ on the other hand. That is, we put

$$\frac{\partial}{\partial t} W = -\nabla \cdot \mathbf{J}_{\mathrm{th}} \, . \tag{17}$$

By means of Eq. (17), the left-hand side of Eq. (16) can be solved formally:

$$\frac{\mathrm{d}}{\mathrm{d}t} F = \int \frac{\delta F}{\delta W} \frac{\partial}{\partial t} W \, \mathrm{d}^N x \tag{18}$$

$$= -\int \frac{\delta F}{\delta W} \nabla \cdot \mathbf{J}_{\mathrm{th}} \, \mathrm{d}^N x \tag{19}$$

$$= \int \mathbf{J}_{\mathrm{th}} \cdot \nabla \frac{\delta F}{\delta W} \, \mathrm{d}^N x \, , \tag{20}$$

where $\delta F / \delta W$ is the so-called variational derivative of F with respect to the probability density W (for details see e.g. Ref. [18]). Comparing Eqs. (16) and (20), we identify the thermodynamical force $\mathbf{X}_{\mathrm{th}}$ as

$$\mathbf{X}_{\mathrm{th}} = -\nabla \frac{\delta F}{\delta W} \, . \tag{21}$$

This is the gradient force of a potential μ:

$$\mathbf{X}_{\mathrm{th}} = -\nabla \mu \, , \quad \mu = \frac{\delta F}{\delta W} \, . \tag{22}$$

We are left to identify $\mathbf{J}_{\mathrm{th}}$. In line with linear non-equilibrium thermodynamics, it is assumed that the flux $\mathbf{J}_{\mathrm{th}}$ is driven by the force $\mathbf{X}_{\mathrm{th}}$ such that $\mathbf{J}_{\mathrm{th}} = \mathbf{G}(\mathbf{X}_{\mathrm{th}})$, where $\mathbf{G}$ is a general function. For zero force, the flux is assumed to be equal to zero as well: $\mathbf{G}(0) = 0$. For weak forces the function $\mathbf{G}$ can be linearized at $\mathbf{X} = 0$ such that there is a linear relationship between $\mathbf{J}_{\mathrm{th}}$ and $\mathbf{X}_{\mathrm{th}}$. This is the domain in which 'linear' non-equilibrium thermodynamic holds. Assuming this linear relationship, we put $\mathbf{J}_{\mathrm{th}} = A \cdot \mathbf{X}_{\mathrm{th}}$, where A is a $N \times N$ matrix. Comparing $\mathbf{J}_{\mathrm{th}} = A \cdot \mathbf{X}_{\mathrm{th}}$ with Eq. (9), we split A into two parts, a $N \times N$ matrix M and the probability density W, such that $A = M\, W$. In summary, the thermodynamic flux is found as [9, 15, 16, 18]

$$\mathbf{J}_{\mathrm{th}} = M \cdot \mathbf{X}_{\mathrm{th}} \, W \, . \tag{23}$$

Substituting Eq. (23) into Eq. (17), we obtain an evolution equation for W of the form

$$\frac{\partial}{\partial t} W = -\nabla \cdot \mathbf{J}_{\mathrm{th}} = -\nabla \cdot M \cdot \mathbf{X}_{\mathrm{th}} \, W \, . \tag{24}$$

However, Eq. (24) neglects the contribution of the conservative force $\mathbf{I}$ as described by Eq. (8). In order to address this contribution, the right-hand side term of Eq. (8) needs to be added to the right-hand side of Eq. (24):

$$\frac{\partial}{\partial t}W = -\nabla \cdot (\mathbf{J}_I + \mathbf{J}_{\text{th}}) = -\nabla \cdot \{\mathbf{I}\,W + M \cdot \mathbf{X}_{\text{th}}\,W\} \ . \tag{25}$$

More explicitly, we have

$$\frac{\partial}{\partial t}W = -\nabla \cdot \mathbf{I}\,W + \nabla \cdot \left\{ M \cdot W\,\nabla \frac{\delta F}{\delta W} \right\} \ . \tag{26}$$

Equation (26) is our final result. Equation (26) describes the evolution of the probability density W. Formally, Eq. (26) exhibits the structure of a Fokker-Planck equation [15, 16, 18, 28, 57]. The trajectories or realizations of a stochastic process described by a Fokker-Planck equation can be computed from a Langevin equation [15, 16, 18, 28, 57]. Consequently, the functions $\mathbf{h}$ and g in Eq. (1) can be identified. Substituting $F = U - QS$ into (26) and comparing Eqs. (1) and (26), we obtain

$$\mathbf{h} = \mathbf{I} - M \cdot \nabla \frac{\delta U}{\delta W} \tag{27}$$

and

$$\sum_{j=1}^{N} g_{ij} g_{kj} = Q M_{ik} \ . \tag{28}$$

Let us point out the benefit of this approach. Let us assume stochastic motor behavior can be described in terms of the energy U and the Langevin equation (1) with coefficients $\mathbf{h}$ and g defined by Eqs. (27) and (28). Then the process energy U as well as the process entropy S and energy F can be studied under different experimental conditions. The measures U, S, and F have important interpretations in the context of motor behavior, see Table 1. Therefore, the variations in U, S, and F observed across experimental conditions provide certain pieces of information about how the motor behavior and the human motor control system is affected by the manipulations at hand. These pieces of information can hardly be obtained otherwise.

2.5. Free Energy Minimization Principle

Finally, let us highlight the link to the Landau theory. Frequently, it can be shown that F is bounded from below [17, 18]. In this case, the model predicts that the motor behavior under consideration eventually becomes stationary in the sense that the probability density $W(\mathbf{x})$ of the motor pattern becomes stationary. The free energy in this stationary regime assumes a minimum. That is, for any probability density W' different from the aforementioned stationary distributions W the free energy F increases.

In the following two sections we will exploit the thermodynamical approach in order to study single limb coordination with a pacing signal and between-persons coordination.

3. Single Limb Coordination with a Rhythmic Pacing Signal: Canonical-Dissipative Coordination Thermodynamics

3.1. Rhythmic Single Limb Coordination with a Pacing Signal as a Benchmark Paradigm of Motor Coordination

Rhythmic movements are part of human every day activities. For example, walking, running, and stair climbing involves rhythmic leg movements. As such rhythmic movements represent a fundamental type of motor activity. In a laboratory setting properties of the control of such movements can be studied most conveniently using a pacing paradigm, where the movement has to be synchronized with the beat of a metronome. In doing so, the movement frequency can be manipulated.

The pendulum swinging paradigm [42] provides a controllable experimental setup for studying this sort of behavior. In this paradigm, the main task is to swing a pendulum about an axis through the wrist at a given frequency. The paradigm has been exploited in various studies. In particular, it has been found that when oscillation amplitude was unconstrained, then the amplitude decayed with the increase of pacing frequency [3, 38, 66]. Moreover, the energy measured on the basis of appropriate oscillator models increased as a function of pacing frequency [37, 43].

3.2. Modeling and Data Analysis

In what follows, we will report data from a study by Dotov and Frank [11] in which paced uni-manual pendulum swinging was investigated. The stochastic oscillatory behavior was modeled in terms of a so-called canonical-dissipative oscillator. The model is a special case of the thermodynamical model introduced in Section 2. For details about canonical-dissipative systems the reader is referred to Refs. [12, 30, 31, 32]. The canonical-dissipative oscillator in particular has been discussed in several previous studies [11, 14, 20, 23] and the connection between canonical-dissipative systems and the so-called W-method for investigating uni-manual rhythmic motor activities [4] has been pointed out as well [11].

According to the canonical-dissipative approach, the oscillation is described by a position variable x and a velocity variable v. The unperturbed oscillatory dynamics is assumed to correspond at least approximately to the dynamics of a harmonic oscillator (3) with $k = \omega^2$, where $\omega > 0$ is the angular frequence $\omega = 2\pi f_{\mathrm{osc}}$ and f_{osc} is the oscillation frequency in Hertz. Consequently, the conservative dynamics satisfies the Hamiltonian oscillator equations

$$\frac{\mathrm{d}}{\mathrm{d}t}\begin{pmatrix} x_r \\ v_r \end{pmatrix} = \mathbf{I} = \begin{pmatrix} v_r \\ -\omega^2 x_r \end{pmatrix} \tag{29}$$

and involves the Hamiltonian energy function

$$H = \frac{v_r^2}{2} + \omega^2 \frac{x_r^2}{2} \tag{30}$$

introduced earlier in Section 2. There are non-conservative forces $F_{\mathrm{pump/losses}}$ that reflect energy pumping and energy losses. In line with Newtonian dynamics, these forces result in acceleration and de-acceleration of the oscillatory motion, that is, in a change of the velocity v. Eq. (29) becomes

$$\frac{\mathrm{d}}{\mathrm{d}t}\begin{pmatrix} x_r \\ v_r \end{pmatrix} = \begin{pmatrix} v_r \\ -\omega^2\, x_r + F_{\text{pump/losses}} \end{pmatrix} . \tag{31}$$

Finally, in order to account for the typically erratic nature of human rhythmic movements, the model is supplemented with a Langevin force $g \cdot \Gamma_r$ as introduced in Sec. 2.. Here $g > 0$ is a scalar. Eq. (31) becomes

$$\frac{\mathrm{d}}{\mathrm{d}t}\begin{pmatrix} x_r \\ v_r \end{pmatrix} = \begin{pmatrix} v_r \\ -\omega^2\, x_r + F_{\text{pump/losses}} + g\Gamma_r(t) \end{pmatrix} . \tag{32}$$

The canonical-dissipative approach states that the non-conservative forces affecting the energy balance are related to the conservative (canonical force) described by the Hamiltonian function. Accordingly, we put $F_{\text{pump/losses}} = -\gamma v_r(H - B)$, where $\gamma > 0$ and B are two model parameters such that

$$\frac{\mathrm{d}}{\mathrm{d}t}\begin{pmatrix} x_r \\ v_r \end{pmatrix} = \begin{pmatrix} v_r \\ -\omega^2\, x_r - \gamma v_r[H(x_r, v_r) - B] + g\Gamma_r(t) \end{pmatrix} . \tag{33}$$

Eq. (32) is the canonical-dissipative oscillator model. It describes the emergence of an oscillatory behavior due to a Hopf-bifurcation. The control parameter is B. For $B < 0$ the deterministic model ($g = 0$) exhibits a stable fixed point at the origin $x = v = 0$. For $B > 0$ the origin is an unstable fixed point and the model exhibits a stable limit cycle describing an oscillatory dynamics. The control parameter B has an alternative interpretation. In the deterministic case, for $B > 0$ the model energy H of a trajectory equals B after a transient period. That is, not only does B represent a control parameter, B also corresponds to the stationary energy $H(st)$ of the deterministic model. The parameter γ defines a time scale on which the energy converges to the stationary energy value [11, 20]. As far as the energy dynamics is concerned that parameter γ is a global measure for the fixed point attractor located at $H = B$. It can be shown that the model exhibits a second stability parameter $2\gamma B$. The product $2\gamma B$ measures attractor strength against small energy perturbations out of the fixed point energy $H = B$ provided $B > 0$, see Ref. [23].

Equation (33) is compatible with the thermodynamical approach sketched in Fig. 1 and described in detail in Sec. 2. That is, Eq. (33) satisfies the form (28) and (29). A detailed calculation shows that $E(\mathbf{x}_r) = E(x_r, v_r) = \gamma[H(x_r, v_r) - B]^2/2$ such that

$$U = \langle E \rangle = \frac{\gamma}{2}\left\langle [H(x_r, v_r] - B)^2 \right\rangle_{x_r, v_r} = \frac{\gamma}{2}\int\int (H(x, v) - B)^2 W(x, v)\, \mathrm{d}x\, \mathrm{d}v . \tag{34}$$

Likewise for the 2×2 matrix M and the parameter g we obtain

$$M = \begin{pmatrix} 0 & 0 \\ 0 & 1 \end{pmatrix} \tag{35}$$

and

$$g^2 = Q . \tag{36}$$

An analytical expression for the stationary probability density $W(\mathbf{x}) = W(x, v)$ can be derived and reads [20]

$$W(x, v) = \frac{1}{Z_{xv}} \exp\left\{-\frac{\gamma}{2Q}[H(x, v) - B]^2\right\} , \tag{37}$$

where $Z_{xv} > 0$ is a normalization constant. It can be shown that in the energy space the energy values $H_r = H(x_r, v_r)$ exhibit a similar distribution [48]

$$P(H) = \frac{1}{Z_H} \exp\left\{-\frac{\gamma}{2Q}[H - B]^2\right\} , \tag{38}$$

where $Z_H > 0$ is another normalization constant. The constants Z_{xv} and Z_H are related to each other by $Z_{xv} = \omega Z_H/(2\pi)$, see Ref. [48]. Note that $P(H)$ is only defined for $H \geq 0$. Consequently, $P(H)$ corresponds to a truncated normal distribution.

Exploiting the definitions for S, F and U given by Eqs. (6), (13), (34), and taking the explicit forms of the distributions (37) and (38) into account, a detailed calculation shows that

$$\begin{aligned} U_{xv} &= \frac{Q}{2}\left(1 - \frac{B}{Z_H}\exp\{\beta B^2/2\}\right) , && (39) \\ S_{xv} &= \ln Z_{xv} + \frac{U_{xv}}{Q} \\ &= \ln Z_H + \ln\left(\frac{2\pi}{\omega}\right) + \frac{U_{xv}}{Q} \\ &= \ln Z_H + \ln\left(\frac{2\pi}{\omega}\right) + \frac{1}{2}\left(1 - \frac{B}{Z_H}\exp\{\beta B^2/2\}\right) , && (40) \\ F_{xv} &= -Q\ln Z_{xv} = -Q\ln Z_H + \ln\left(\frac{2\pi}{\omega}\right) && (41) \end{aligned}$$

with

$$Z_H = \frac{2\pi}{\beta}\left(1 - \Phi(-B\sqrt{\beta})\right) \tag{42}$$

and

$$\beta = \frac{\gamma}{Q} . \tag{43}$$

In Eq. (42) the function Φ is the so-called error function (or cumulative probability of the normal distribution): $\Phi(z) = \int_{-\infty}^{z} \exp\{-y^2/2\}\,\mathrm{d}y$. We added the sub-index xv to the thermodynamical variables in order to indicate that they are computed from the probability density $W(x, v)$ defined on the two-dimensional phase space spanned by the coordinates x and v. Analogous thermodynamical variables can be defined in the energy space spanned by the energy variable $H \geq 0$. More precisely, we have

$$\begin{aligned} U_H &= \frac{\gamma}{2}\int_0^\infty (H - B)^2\, P(H)\,\mathrm{d}H , && (44) \\ S_H &= -\int_0^\infty P(H)\,\ln P(H)\,\mathrm{d}H , && (45) \\ F_H &= U_H - QS_H . && (46) \end{aligned}$$

A detailed calculation shows that the internal energy is the same for phase space and energy space considerations. Likewise, S_H and F_H correspond to S_{xv} and F_{xv} but do not include the additive ω-term. That is, we have

$$U_H = U_{xv} , \tag{47}$$

$$\begin{aligned} S_H &= \ln Z_H + \frac{U_H}{Q} \\ &= \ln Z_H + \frac{1}{2}\left(1 - \frac{B}{Z_H}\exp\{\beta B^2/2\}\right) , \end{aligned} \tag{48}$$

$$F_H = -Q \ln Z_H . \tag{49}$$

The model parameters and thermodynamical variables as function of the model parameters are summarized in Table 2.

The questions arises how do the model parameters and thermodynamical variables vary with the pacing frequency ω of the performed rhythmic coordination task. Since the Hamiltonian energy H depends on ω^2 rather than ω, see Eq. (30), it is more plausible to vary ω^2 and to examine the variations of model parameters and thermodynamic variables across different values of ω^2.

Table 2. Model parameters and thermodynamical variables of the canonical-dissipative description of rhythmic uni-manual motor activity. The index xv/H at S and F indicates that one should distinguish between two subtypes of thermodynamical variables: S_{xv} versus S_H and F_{xv} versus F_H. For U both subtypes assume identical function values.

Model parameters				Thermodynamic variables		
Energy pump.	Attractor strength		Fluct. strength	Energy		Entropy
	Global	Local		Internal	Free	
B	γ	$\lambda^* = 2\gamma B$	Q	$U(\gamma, B, Q)$	$F_{xv/H}(\gamma, B, Q)$	$S_{xv/H}(\gamma, B, Q)$

3.3. Dotov and Frank (2011) Revisited: Methods and Results

3.3.1. Experiment and Analysis

Participants Six undergraduate students from the University of Connecticut participated in exchange for course credit. All experimental procedures were approved by the Institutional Review Board of the University of Connecticut.

Material and Apparatus A goniometer centered about the axis of radial excursion of the wrist measured the wrist angle at a sampling frequency of 100 Hz and stored the digitized data on a computer. The pendulum consisted of a metal rod with a wooden handle on one end and a cylindrical weight on the other, measured 40 cm in length and 297.5 grams in weight. For the rod-hand system this yielded an equivalent pendulum with a resonance frequency of 6.07 rad$/s$ or .97 Hz. The pendulum was kept the same across trials but an auditory pacing signal was used to set the participant's swinging at different frequencies. A chair with a specially designed right-hand side arm-rest was used.

Procedure Prior to data collection the participants were familiarized with the apparatus and the task, observed the experimenter perform the task, and then performed a practice trial. Participants were instructed to synchronize the anterior extrema of the wrist excursions with the auditory pacing signal. The importance of the acoustic pacing signal and that all motions were to be performed about a single axis in wrist was stressed in the instructions. After that participants performed eighteen one-minute-long trials with short breaks in between.

Design and Analysis The frequencies were chosen randomly from among five possible frequencies (cf. Table 3). Three trials per frequency were performed. Three self-paced trials were interspersed among the other trials but were not used in the analysis. Descriptive measures and parameter estimation per participant was performed for each trial. In addition to the model-based data analysis described below, performance frequency and performance amplitude were estimated for each trail (for details see Ref. [11]). For data reduction purposes all measures were averaged within the three repeating trials in given frequency. Repeated measures analysis of variance was performed to test for effects of pacing frequency. The significance level was set at 0.05. The lowest frequency condition was dropped from the analysis because under this pacing condition qualitatively different types of trajectories were observed and, accordingly, very different measures which demand separate treatment.

Model-based data analysis From experimental data given in term of trajectories $x_r(t)$, velocity trajectories $v_r(t)$ were constructed and eventually energy values H_r were estimated [20, 11]. Moreover by means of Eq. (38) and the trajectories $x_r(t)$ and $v_r(t)$, estimates $\hat{B}$, $\hat{\gamma}$, and $\hat{Q}$ for the model parameters B, γ, and Q were obtained following the protocol described in Refs. [11, 20, 25]. Given estimates for B, γ, and Q, the thermodynamical variables U, S, and F were computed from Eqs. (39)-(43) and (48), (49).

3.3.2. Results as Reported in Dotov-Frank-2011

Figures 3 shows the performed frequencies plotted versus the pacing frequencies. Overall, participants performed pendulum swinging at the required frequencies. The squared amplitudes of the observed rhythmic wrist movements are reported in Table 3 for the different pacing frequencies. For the six participants (N = 6) the mean of the squared amplitude decreased with frequency increase. The model parameters B, Q, γ, and λ^* are shown in Table 4. It was found that B increased, Q appeared not to change, γ decreased and λ^* increased. λ^* approximated $2\gamma < H >$ as calculated in Dotov and Frank [11]. The main effect on squared amplitude was significant, F(3,15)=3.34, $p < .05$, with a decreasing linear trend ($p < .05$). The effect on B was significant, F(3,15)=8.39, $p < .01$, with an increasing linear trend ($p < .01$). Q showed only a marginally significant effect, F(3,15)=3.01, $p = .06$. The effect on γ was significant, F(3,15)=7.47, $p < .05$, with an increasing linear trend ($p < .05$). The effect on λ^* was not significant.

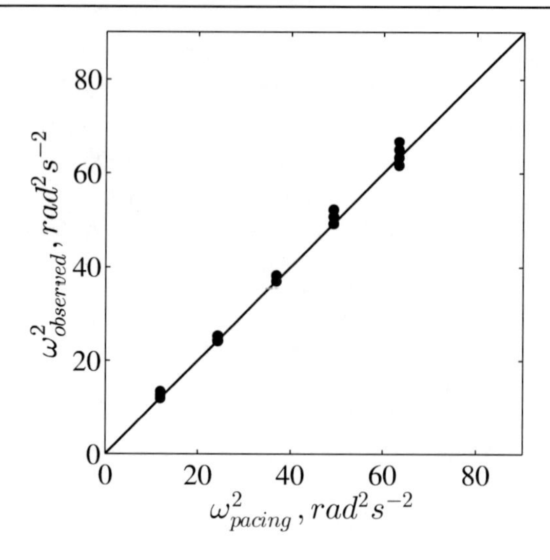

Figure 3. Performed frequency versus pacing frequency

Table 3. Squared amplitude as a function of the squared pacing frequency (as reported in [11]) with SDs in parentheses

ω^2	11.94	24.22	36.89	49.23	63.34
Squared Amplitude	0.17 (0.07)	0.15 (0.07)	0.14 (0.07)	0.12 (0.05)	0.11 (0.05)

3.3.3. Results from the Thermodynamical, Model-Based Analysis Beyond the Dotov-Frank-2011 Study

Table 4 also presents sample means of the thermodynamical variables. The internal energy variable U seemed to exhibit an inverse quadratic trend, where the top of the curve was around the resonant frequency. S_{xv} and F_{xv} did not change, while S_H appeared to increase in magnitude and F_H decayed as a function of ω^2. The effect on U was only marginally significant, F(3,15)=3.13, $p = .057$. For both S_{xv} and F_{xv} there was no significant effect. The effect of frequency on S_H was statistically significant, F(1.20,5.98)=11.52, $p < .05$, with an increasing trend ($p < .05$). The effect on F_H was statistically significant, F(3,15)=4.56, $p < .02$, with the trend in the negative direction marginally significant at $p = 0.06$.

Table 4. Model parameters and thermodynamical variables obtained for different pacing conditions. Note that $\lambda^* = 2\gamma B$ only holds for $B > 0$. Since in the very slow pacing condition negative values of B were observed, the definition does not apply in this condition.

	Model parameters				Thermodynamic variables				
	Energy pumping	Attractor strength Global	Attractor strength Local	Fluct. strength	Energy Internal	Energy Free	Energy Free	Entropy	Entropy
ω^2	B	γ	λ^*	Q	$U_{xv} = U_H$	F_{xv}	F_H	S_{xv}	S_H
11.94	−0.4 (2.0)	24 (11)	N.A.	45 (17)	56 (63)	−20 (51)	7 (57)	1.7 (.3)	1.1 (.3)
24.22	1.9 (1.0)	23 (10)	73 (30)	37 (13)	15 (5)	−46 (17)	−37 (15)	1.7 (.3)	1.5 (.3)
36.89	2.7 (1.9)	17 (9)	82 (28)	41 (14)	19 (5)	−53 (22)	−52 (22)	1.8 (.3)	1.7 (.3)
49.23	3.1 (1.7)	17 (9)	88 (32)	46 (14)	21 (6)	−58 (28)	−63 (21)	1.7 (.3)	1.8 (.3)
63.34	4.0 (1.7)	13 (7)	88 (20)	41 (17)	18 (6)	−53 (30)	−63 (34)	1.7 (.3)	1.9 (.3)

3.4. Discussion

The pumping parameter B increased with squared pacing frequency. In fact, this is consistent with the literature in which an increase of oscillator energy was reported [37, 43]. The reason for this is that the oscillator energy reported in Ref. [37, 43] reflects $\langle H \rangle$ (and is not related to either of the two thermodynamical energy variables U and F) and $\langle H \rangle$ in turn for vanishing fluctuations would equal B (see Sec. 3.2.).

As reported earlier in Dotov and Frank [11], the parameter γ decreased, whereas λ^* increased initially and reached a plateau. Dotov and Frank argued that this was consistent with results reported in Ref. [4].

The internal energy U increased in the three frequency plateaus $\omega^2 = 24.22, 36.89$ and 49.23 and finally dropped. This pattern gives rise to the speculation of a quadratic trend with a peak at the resonance frequency of the pendulum. This hypothesis seems to be worth being investigated in more detail in future experimental studies.

Most importantly, we found that free energy F_H decreased. Figure 4 depicts F_H as function of the squared pacing frequencies. The more or less monotonically decaying pattern is consistent with the Landau theory of phase transitions in equilibrium systems (see Fig. 2) when assuming that for higher pacing frequencies ω^2 the motor control system is taken further and further away from the Hopf bifurcation point. This assumption, in turn, is supported by the observed increase of the control parameter B with increasing pacing frequencies ω^2.

In addition, we would like to point out an alternative interpretation of the precise value of F_H. The frequency conditions were presented in random order (see Sec. 3.3.1.). Therefore, we may assume that *on the average* the motor control systems involved in the rhythmic performance task were on the same initial free energy level F_0 for all pacing conditions. In this case, the observed free energy values F_H represents energy shifts ΔF from the unknown baseline value F_0 to the observed value F_H. The arrows in Fig. 4 indicates these energy shifts. The consideration of energy shifts is highly important because usually absolute energy values are not of primary interest. What typically matters are energy gradients that induce forces. In our case, the fact that the free energy shifts ΔF became larger when squared pacing frequency ω^2 were increased is regarded as a key observation.

The entropy S_H increased. Moreover, the measures S_{xv} and F_{xv} were not affected by the pacing frequency. We will leave it to future studies to reflect on the meaning of the differences between S_{xv} and F_{xv} and S_H and F_H. In contrast, we would like to address the issue of an entropy increase S_H. However, we will do so in Section 5., where we also address the entropy change observed in the between-persons coordination experiment.

Finally, Figure 5 presents in a two dimensional plane the model parameters γ and B as obtained under the four higher pacing conditions $\omega^2 = 24.22, 36.89, 49.23$ and 63.34. A more or less linear relationship can be observed. This is somewhat counter intuitive because $\lambda^* = 2\gamma B$ reaches a plateau for the three highest pacing conditions $\omega^2 = 36.89$, 49.23, 63.34. That is, λ^* is almost constant, which would suggest a non-linear, hyperbolic relationship between γ and B. Future work needs to elucidate this point. However, the main purpose of Figure 5 is to illustrate the possibility that the parameters γ and B change in a lawful way as functions of ω^2 such that they may represent a one-dimensional graph in the $\gamma - B$ plane rather than a two-dimensional scattered cloud. In other words, Figure 5

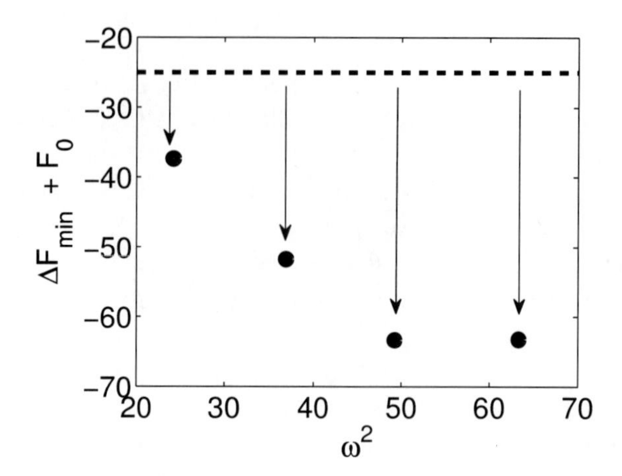

Figure 4. Re-interpretation of $F(\omega^2)$ as $\Delta F_{\rm min} + F_0$, which is the change in minimal free energy shifted by F_0. That is, $\Delta F_{\rm min} + F_0 = F(\omega^2)$ is plotted versus ω^2 for $\omega^2 = 24.22$, 36.89, 49.23, 63.34 [1/s^2] using $F(\omega^2) = F_H(\omega^2)$. F_0 is indicated as dashed 'symbolic' line. The true value of F_0 is unknown. Compare with Fig. 2 (right).

invites us to speculate to what extent motor coordination systems tested in the pendulum swinging paradigm are subjected to constraints such that a change $\Delta\omega^2$ in squared pacing frequency ω^2 induces changes $\Delta\gamma$ and ΔB that are correlated such that $\gamma(\omega^2)$ and $B(\omega^2)$ correspond to a parametric representation of a one-dimensional curve.

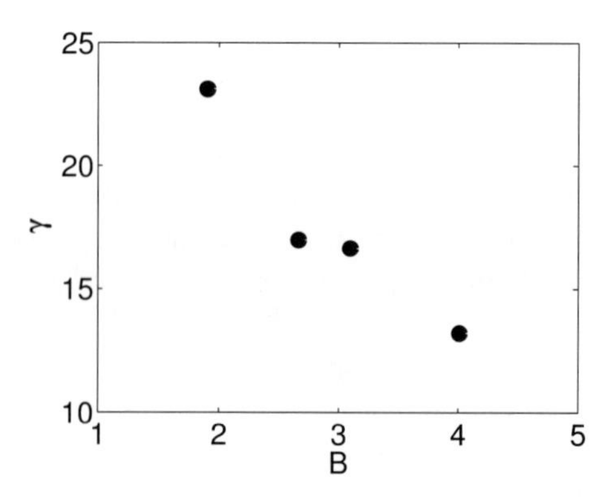

Figure 5. Plot γ versus B for $\omega^2 = 24.22, 36.89, 49.23, 63.34$ [1/s^2]

4. Between Persons Coordination: Mean Field Coordination Thermodynamics

4.1. Small Group Research Between Large Group Motor Coordination and Dyadic Motor Coordination

Coordinated motor behavior in large groups of people typically has been studied in the context of rhythmic synchronized activity [51]. Inspired by Winfree's seminal studies on synchronization of animal behavior [68] (e.g., firefly flashings [6, 7]), synchronized applause [49, 50] and synchronized stepping of pedestrians as cause for the Millennium bridge instability [13, 64] have been investigated as large group coordination phenomena. On the other end of the size scale, synchronized motor coordination between two people has been studied. In particular, factors have been examined that affect the strength of visual coupling that enables two actors to perform synchronized motor activities [58, 59, 60]. Synchronized coordinated activity between pairs of people (dyads) is of particular importance for the social sciences [46, 47]. For example, it has been reported that performance of certain social tasks improves when dyads coordinate their motor activities in a synchronized way [56, 61]. That is, motor coordination impacts social dynamics.

In search for a useful experimental paradigm, the rocking chair paradigm has been developed in a series of studies [54, 55]. In this paradigm, participants are seated in rocking chairs and are asked to rock back and forth. Motor coordination in terms of the synchronization of the rocking movements can be studied under various conditions. Most importantly, the paradigm can be conducted with two people (i.e., as a dyad experiment) or with more than two persons as an experiment in the field of small group research. Small group laboratory research on coordination, in turn, is a relatively newly developing discipline. It is relevant for our understanding of team sports, on the one hand, and its social aspects are important for group decision making, in general, and jury decision making, in particular.

4.2. Modeling I: the Classical Kuramoto Model in the Life Sciences

One of the most fundamental models for understanding phase synchronization in large oscillator populations is the phase oscillator model introduced by Kuramoto [44]. Reviews of this model can be found in Ref. [1, 18, 19, 22, 63, 68].

We consider a set of N interacting phase oscillators. Each oscillator is described by its angle or phase φ_i, $i = 1, \ldots, N$. Without loss of generality, the oscillator phases are defined on the interval $\varphi_i \in [-\pi, \pi]$, $i = 1, \ldots, N$. Individual, isolated oscillators are assumed to exhibit an oscillation frequency f_i such that $\varphi_i = \omega_i t + \varphi_{i,0}$ holds, where $\omega_i = 2\pi f_i$ denotes the angular eigenfrequency and $\varphi_{i,0}$ describes the initial phase at time $t = 0$. In the most fundamental case, the oscillator frequencies are homogeneous, that is, $\omega = \omega_i$ for all $i = 1, \ldots, N$. In what follows, we will restrict ourselves to consider the homogeneous case. In this case, the oscillators can be studied from the perspective of a rotating laboratory system that represents a phase oscillator with oscillation frequency ω. Mathematically speaking, we introduce the shifted phases $x_i(t) = \varphi_i(t) - \omega t$ with $x_{i,0} = \varphi_{i,0}$. In the rotating framework, the dynamics of a phase oscillator as described by $x_i(t)$ is affected by the coupling between oscillators and fluctuating forces acting on the oscillators. The impact of the intrinsic dynamics on the phase oscillators is eliminated. In this context, the evolution equations for $x_{i,r}$ may read like

$$\frac{\mathrm{d}}{\mathrm{d}t} x_{r,j} = -\frac{\kappa}{N} \sum_{k=1,k\neq j}^{N} \sin(x_{r,j} - x_{r,k}) + \sqrt{Q}\,\Gamma_{r,j} \tag{50}$$

with $i = 1, \ldots, N$ and $\kappa\ , Q \geq 0$. The sine-term describes the coupling between the oscillators. The parameter $\kappa \geq 0$ is a measure for the strength of coupling. For $\kappa > 0$ the interaction described by the sine-term is attractive. That is, for vanishing fluctuating force ($Q = 0$), the phases converge to the same angle θ. That is, we have $x_{r,j} = \theta$ for $j = 1, \ldots, N$. Eq. (50) describes the basic version of the Kuramoto model. The Kuramoto model (50) describes a stochastic motor coordination process of the form (1) with

$$h_j = -\frac{\kappa}{N} \sum_{k=1,k\neq j}^{N} \sin(x_{r,j} - x_{r,k})\ . \tag{51}$$

The degree of phase synchronization can be quantify by means of circular statistics. To this end, frequently the cluster amplitude $r \geq 0$ and the cluster phase θ defined by

$$r \exp\{i\theta\} = \frac{1}{N} \sum_{k=1}^{N} \exp\{i x_{r,k}\} \tag{52}$$

have been used. Here, $i = \sqrt{-1}$ is the imaginary unit. The cluster phase is a measure of central tendency in circular statistics (analogous to the mean value in ordinary statistics). In contrast, the cluster amplitude r in our context is a measure for the degree of phase synchronization. r can assume values between zero and one and is maximal (i.e., $r = 1$) if there is perfect phase synchronization (i.e., $x_{r,j} = \theta$ for $j = 1, \ldots, N$). When the phase synchronization is only partial and the frequency distribution of the values $x_{1,r}, \ldots, x_{N,r}$ deviates more and more from the special case of perfect synchronization, then r assumes values smaller than 1 (see also below).

The Kuramoto model has been extensively studied for large many-body system. That is, typically the so-called thermodynamic limit, that is, the case of an infinitely large system ($N \to \infty$) has been studied Ref. [1, 18, 19, 22, 63]. In this case, it is sufficient to consider one representative subsystem x_r with probability density $W(x,t)$. The evolution equation for this subsystem reads

$$\frac{\mathrm{d}}{\mathrm{d}t} x_r = -\kappa r_\infty(t) \sin(x_r - \theta_\infty(t)) + \sqrt{Q}\,\Gamma_r\ . \tag{53}$$

Eq. (53) involves the cluster amplitude r_∞ and cluster phase θ_∞ that are defined in a self-consistent way by

$$r_\infty(t) \exp\{i\theta_\infty(t)\} = \int_{-\pi}^{\pi} \exp\{ix\} W(x,t)\,\mathrm{d}x\ . \tag{54}$$

The model (53) predicts that in the long time limit ($t \to \infty$) the process exhibits a stationary probability density

$$W(x) = \frac{1}{Z} \exp\left\{\beta' r_\infty(st) \cos(x - \theta_\infty(st))\right\}\ , \tag{55}$$

where Z denotes the normalization constant $Z = \int_{-\pi}^{\pi} \exp\{\beta' r_\infty(st)\cos(x - \theta_\infty(st))\}\,\mathrm{d}x$. Here, β' represent the ratio

$$\beta' = \frac{\kappa}{Q}\,. \tag{56}$$

For $r_\infty(st) = 0$ the probability density (55) reduces to the uniform distribution, see Fig. 6. In contrast, for $r_\infty(st) > 0$ the probability density is a unimodal. In this case, the stationary cluster phase $\theta_\infty(st)$ can assume arbitrary values and indicates the mode of the distribution, see Fig. 6 again.

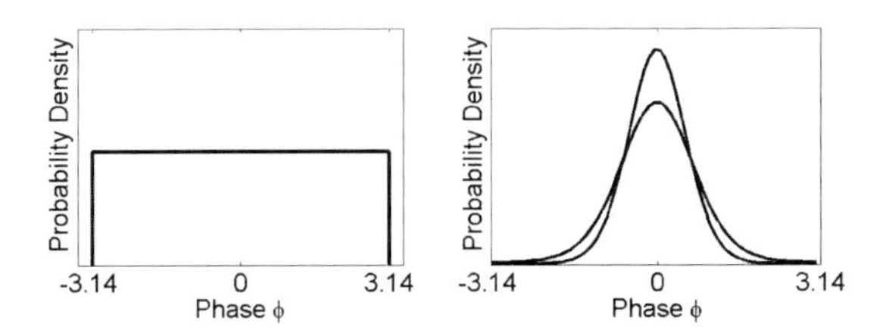

Figure 6. Illustration of the uniform phase distribution (left) and examples of two unimodal phase distributions (right) with $\theta_\infty(st) = 0$.

The cluster amplitude $r_\infty(st)$ is only implicitly defined. The implicit equation reads [18]

$$r_\infty(st) = \frac{1}{Z(r_\infty(st))} \int_{-\pi}^{\pi} \cos(\xi) \exp\{\beta' \, r_\infty(st)\cos(\xi)\}\,\mathrm{d}\xi\,. \tag{57}$$

Equation (57) can be solved numerically. In doing so, the bifurcation diagram shown in Fig. 7 can be computed [18, 26].

Accordingly, the uniform distribution with $r_\infty(st) = 0$ is a solution of the model (53) for any parameters $\kappa, Q \geq 0$. However, only for $\beta' = \kappa/Q < 2$ the uniform distribution is asymptotically stable in the sense that any perturbation in the distribution of phase oscillators will decay. For $\beta' = \kappa/Q > 2$ the uniform distribution is unstable. For $\beta' = \kappa/Q > 2$ there is an infinitely large set of stable distributions (55) that differ with respect to the position of the mode $\theta_\infty(st)$. Starting with a perturbed uniform distribution, the set of phase oscillators will converge to a state described by one of the distributions out of this set of stable unimodal distributions. In doing so, the Kuramoto model (53) describes the emergence of phase synchronization. If the control parameter $\beta' = \kappa/Q$ is increased from sub-threshold

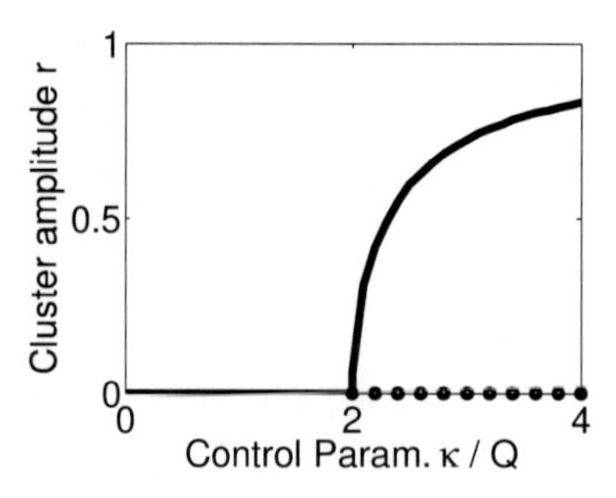

Figure 7. Cluster amplitude $r_\infty(st)$ as function of the control parameter $\beta' = \kappa/Q$. The cluster amplitude $r_\infty(st)$ was computed by solving Eq. (57) numerically.

values smaller than $\beta'_c = 2$ to an above-threshold value larger than $\beta'_c = 2$, then a transition from a non-synchronized phase oscillator ensemble to a partially synchronized ensemble occurs.

Let us turn next to the thermodynamical variables of the infinite dimensional Kuramoto model. We will use lower case letters u, s, and f for internal energy, entropy, and free energy in order to indicate that the thermodynamical variables are variables of a single representative subsystem. That is, they are so-called intensive thermodynamical variables. We need to pay attention to this property because we will consider below the finite dimensional Kuramoto model that involves thermodynamical variables that scale with the number of oscillatory units. The entropy s is defined by Eq. (6). like $s = -\int W \ln W \, dx$. The internal energy u is given by [18]

$$u = -\frac{\kappa}{2} r_\infty(t)^2 \tag{58}$$

and the free energy can be calculated from $f = u - Q\,s$ as in Eq. (13). The Kuramoto model (53) with $f = u - Q\,s$ and s and u defined by Eqs. (6) and (58) corresponds to the Fokker-Planck equation (26) with $\mathbf{I} = 0$ and $M = 1$, see Ref. [18]. That is, the Langevin equation (53) corresponds to the Fokker-Planck equation

$$\frac{\partial}{\partial t} W = \nabla \cdot \left\{ W \cdot \nabla \frac{\delta f}{\delta W} \right\} . \tag{59}$$

In the stationary case ($t \to \infty$) the cluster amplitude and the free energy converge to their respective stationary values: $r_\infty \to r_\infty(st)$, $f \to f(st)$. The stationary free energy is the minimal possible free energy $f(st) = f_{\min}$ (minimal free energy principle, see Sec. 2.5.). From Eqs. (3) and (58) it follows that

$$f_{\min} = -\frac{\kappa}{2} r_\infty(st)^2 - Q\, s(st) . \tag{60}$$

The minimal free energy of the stationary case can be calculated numerically as a function of the control parameter β'. To this end, $r_\infty(st)$ is taken from the bifurcation diagram, see Fig. 7, and $s(st)$ is computed from $s = -\int W \ln W \, dx$ and Eq. (55). Figure 8 shows $f_{\min}$ as function of β' thus obtained. We see that $f_{\min}$ decays as function of the control parameter, which is consistent with the Landau theory reviewed in Sec. 2. (cf. with Fig.2, right).

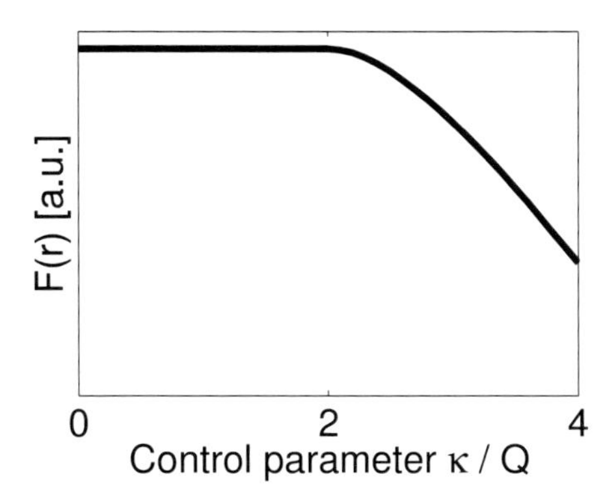

Figure 8. Free energy $f_{\min}$ defined by Eq. (60) as function of β'.

4.3. Modeling II: the Finite Dimensional Kuramoto Model

Recently, the finite dimensional Kuramoto model has been exploited in small group research on motor control to describe between-person coordination [27, 54]. Let us briefly sketch the thermodynamical perspective of this N-dimensional model.

For $N < \infty$ Eq. (50) can equivalently be expressed as

$$\frac{\mathrm{d}}{\mathrm{d}t} x_{r,j} = -\frac{\partial}{\partial x_{r,j}} E(\mathbf{x}_r) + \sqrt{Q}\,\Gamma_{r,j} \tag{61}$$

with

$$E(\mathbf{x}_r) = -\frac{\kappa}{2N} \sum_{j,k=1,j\neq k}^{N} \cos(x_{r,j} - x_{r,k}) \,. \tag{62}$$

The function E can be regarded as an energy function of the phase oscillators. In the deterministic case ($Q = 0$), the energy E is minimal for perfect synchronization (i.e., for $x_{r,k} = \theta$ for all k). The minimal value is given by

$$E_{\min} = -\frac{\kappa}{2}(N-1) \,. \tag{63}$$

In the deterministic case, for arbitrary initial conditions the phase dynamics converges to perfect synchronization such that $E \rightarrow E_{\min}$. In the general case, that is, if $Q > 0$ holds, for $t \rightarrow \infty$ the coordination behavior becomes stationary. The stationary probability density reads [57]

$$W(\mathbf{x}) = \frac{1}{Z_N} \exp\left\{-\frac{E(\mathbf{x})}{Q}\right\} , \tag{64}$$

where Z_N is the normalization constant given by

$$Z_N = \int_{-\pi}^{\pi} \cdots \int_{-\pi}^{\pi} \exp\left\{-\frac{E(\mathbf{x})}{Q}\right\} \mathrm{d}^N x \,. \tag{65}$$

In particular, if there is no coupling between the oscillators ($\kappa = 0$), then the set of oscillators is de-synchronized. In this case, the probability density (64) becomes the uniform

N-dimensional distribution of phases:

$$W(\mathbf{x}) = \left(\frac{1}{2\pi}\right)^N . \tag{66}$$

The degree of synchronization may be quantified by means of the cluster amplitude r defined by (52). However, for hypothesis testing it turns out that the squared amplitude y is a more appropriate measure [27]. The squared amplitude is defined by [27]

$$y(\mathbf{x}) = [r(\mathbf{x})]^2 = \frac{1}{N^2}\left(\sum_{k=1}^{N}\exp\{ix_{r,k}\}\right)\left(\sum_{k=1}^{N}\exp\{-ix_{r,k}\}\right) . \tag{67}$$

More explicitly, we have

$$\begin{aligned} y(\mathbf{x}) &= \frac{1}{N^2}\left\{N + \sum_{j,k=1,j\neq k}^{N} \cos(x_j - x_k)\right\} \\ &= \frac{1}{N^2}\left\{N + 2\sum_{j=1}^{N}\sum_{k=j+1}^{N} \cos(x_j - x_k)\right\} . \end{aligned} \tag{68}$$

The expected squared cluster amplitude $\langle y\rangle$ as function of the control parameter $\beta' = \kappa/Q$ is formally given by

$$\beta' \rightarrow \langle y\rangle = \int_{-\pi}^{\pi}\cdots\int_{-\pi}^{\pi} y(\mathbf{x})W(\mathbf{x};\beta')\,\mathrm{d}^N x . \tag{69}$$

Figure 9 illustrates the graph $\langle y\rangle\,(\beta')$ for $N = 5$. For any given N, the graph $\langle y\rangle\,(\beta')$ can be used as calibration curve to estimate β' from experimentally observed estimates of $\langle y\rangle$.

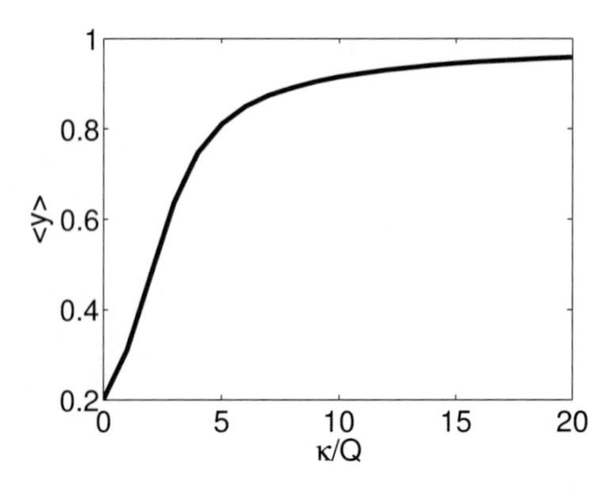

Figure 9. Graph of the mean squared cluster amplitude $\langle y\rangle$ versus the control parameter β' as computed numerically from Eq. (69).

Finally, let us consider the thermodynamical variables U, S, F of the finite dimensional Kuramoto model. In general, these variables depend on the system size N (they are so-called extensive variables). Therefore, we will use the notation U_N, S_N, F_N. Formally, the entropy is defined by the N-dimensional integral

$$S_N = \int_{-\pi}^{\pi} \cdots \int_{-\pi}^{\pi} W(\mathbf{x}, \beta') \ln W(\mathbf{x}; \beta') \, \mathrm{d}^N x \tag{70}$$

with W given by Eq. (64). Likewise, the internal energy U corresponds to the average over E and can be computed from the integral

$$U_N = \langle E \rangle = \int_{-\pi}^{\pi} \cdots \int_{-\pi}^{\pi} E(\mathbf{x}) W(\mathbf{x}; \beta') \, \mathrm{d}^N x \, . \tag{71}$$

Finally, the free energy F_N can be calculated from

$$F_N = U_N - Q\, S_N \, . \tag{72}$$

Technically speaking, it is useful to compute U_N first. Subsequently, S_N and F_N can be computed from

$$S_N = \ln(Z_N) + \frac{U_N}{Q} \, , \tag{73}$$

$$F_N = -Q \ln(Z_N) \tag{74}$$

Figure 10 illustrates U_N, S_N, and F_N as functions of β' for a given system size N.

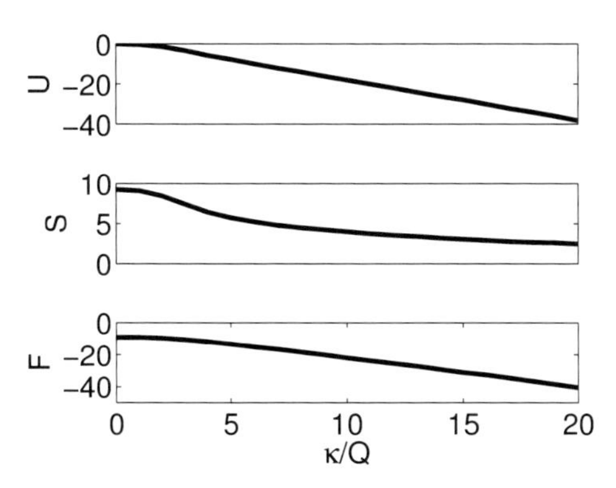

Figure 10. Illustration of U_N, S_N, and F_N as functions of β' as computed from Eqs. (62), (64), (65), (71), (73), and (74). Note that S_N only depends on β'. U_N depends both on $\beta' = \kappa/Q$ and κ. Similarly, F_N depends both on $\beta' = \kappa/Q$ and Q. Therefore, for calculating U_N we fixed $\kappa = 1$ and varied β' (which means that effectively $1/Q$ was varied). For calculating F_N we fixed $Q = 1$ and varied β' (which means that effectively κ was varied.

4.4. Frank and Richardson (2010) Revisited: Methods and Results

4.4.1. Experiment and Analysis

Participants Six groups with six participants each participated in this experiment. Participants were Colby College undergraduate students with no known history of motor disabilities.

Materials The rocking chair experiment (see Sec. 4.1.) involved six wooden rocking chairs. The chairs were arranged in a circle with a radius of 1.25 m around a target in the middle of the circle. The target was positioned at a hight of 1.2 m. The up-down movements of the rocking chairs were recorded with a recording frequency of 120 Hz using a magnetic tracking system.

Design and Analysis In the study by Frank and Richardson [27], three interaction conditions were tested. In this review, only two of them will be considered: 'eyes closed' and 'intentional'. All groups performed 'eyes closed' first and subsequently two repetitions of the 'intentional' condition were performed. Each trial took 3 min. From up-down position data of the chairs, velocity data were calculated. Subsequently, running phases were calculated from the trajectories projected in the two-dimensional position-velocity space (see [27] for details). Running phases were transformed to phases in a rotating (moving) framework. The squared cluster amplitude y was calculated from Eq. (67) using the phases in the rotating framework. The mean squared cluster amplitude $\langle y \rangle$ was computed by replacing ensemble averaging with time-averaging across a trial. In doing so, for each trial a score $\langle y \rangle_T$ was obtained.

Procedure Participants were randomly assigned to the chairs. Participants were given a 1 min practice session in which they were asked to rock in the chair at a comfortable rocking frequency having their eyes closed. Subsequently, participants performed in three experimental interaction conditions. Here we consider only two of them: the 'eyes closed' and 'intentional' conditions. In the 'eyes closed' interaction condition, participants were asked to rock back and forward at a comfortable speed with their eyes closed for a duration of 3 min. In the 'intentional' condition, participants were asked to rock back and forward at the comfortable speed for a period of 3 min while looking at the central target. They were asked to synchronize their rocking movements with the other members of the group while looking at the target.

Model-based data analysis beyond the Frank-Richardson-2010 study In the original study by Frank and Richardson [27], no attempts were made to analyze the data with respect to the stochastic, thermodynamical Kuramoto model (50). For the present study, such a model-based analysis was conduction.

One sensor was broken. Consequently, a five phase oscillator model ($N = 5$) was used to conduct the data analysis based on the thermodynamical, stochastic finite-dimensional Kuramoto model. Ergodicity was assumed to hold. The time-averaged measure $\langle y \rangle_T$ was considered as equivalent to an ensemble averaged measure $\langle y \rangle$. From the calibration graph shown in Fig. 9, the control parameter β' was estimated for each given score $\langle y \rangle_T$. Subsequently, the entropy score S_N was determined from the graph shown in Fig. 10 (middle panel). Moreover, the following two rescaled thermodynamical variables were examined

$$\left(\frac{U_N}{\kappa}\right)(\beta') = \frac{\langle E \rangle}{\kappa} = \int_{-\pi}^{\pi} \cdots \int_{-\pi}^{\pi} \frac{E(\mathbf{x})}{\kappa} W(\mathbf{x};\beta')\, \mathrm{d}^N x \,, \tag{75}$$

$$\left(\frac{F_N}{Q}\right)(\beta') = -\ln(Z_N(\beta'))\,. \tag{76}$$

As indicated, both rescaled thermodynamical variables depend only on β'. In principle, the rescaled thermodynamical variables U_N/κ and F_N/Q were computed by taking the graphs shown in Fig. 10 (top and bottom panels) and rescaling them appropriately. In practice, however, it was noted that in Fig. 10 (top panel) the graph U_N was computed for $\kappa = 1$, which was equivalent to plotting U_N/κ over β'. Likewise, the graph F_Q (bottom panel) was computed for $Q = 1$, which was equivalent to plotting F_N/Q over β'. Consequently, for pragmatic reasons, the graphs for U_N and F_N shown in Fig. 10 were used to determine U_N/κ and F_N/Q for given estimates of β'.

4.4.2. Results as Reported in Frank-Richardson-2010

Table 5 shows $\langle y\rangle_T$ reported in Ref. [27] (as taken from Table 2 of Ref. [27]). In view of Fig. 9, under the null hypothesis $\beta' = 0$ it was expected that $\langle y\rangle = 0.2$ holds (see also Ref. [27] for an alternative motivation of this argument). For the 'eyes closed' interaction condition, observed squared cluster amplitudes were close to this value, whereas in the 'intentional' interaction condition values were found that were much larger than $\langle y\rangle = 0.2$.

Table 5. Time-averaged squared cluster amplitudes $\langle y\rangle_T$ reported in Ref [27] for six test groups tested under two interaction conditions.

Group	Eyes closed	Intentional		
		Rep. 1	Rep. 2	Av.(1+2)
1	0.21	0.17	0.56	0.37
2	0.24	0.42	0.30	0.36
3	0.20	0.72	0.62	0.67
4	0.20	0.56	0.65	0.61
5	0.20	0.75	0.73	0.74
6	0.20	0.71	0.66	0.69

Table 6. Sample means (and SDs) of the scores presented in Table 5.

Condition	Sample Mean $M_{\langle y\rangle_T}$
Eyes closed	0.21 (0.02)
Intentional	0.57 (0.17)

Table 6 shows the sample mean values of the scores shown in Table 5 (cf. also Table 4 in Ref. [27]). The sample mean of the 'eyes closed' condition was $M = 0.21$. Taken the two repetitions together (see the averaged scores in Table 5), the sample mean of the 'intentional' condition was $M = 0.57$. Frank and Richardson [27] reported that the sample mean value of $M = 0.21$ of the 'eyes closed' interaction condition was not significantly different from the test value of 0.2 (t test, one-tailed, significance level 0.05). In contrast, when evaluating the averaged scores for the 'intentional' interaction condition, then the sample mean value of $M = 0.57$ was statistically different from the test value of 0.2 ($t(5) = 5.47$ with $t_c(5) = 2.02$ at a significance level of 0.05).

4.4.3. Results from the Thermodynamical Model-Based Analysis Beyond the Frank-Richardson-2010 Study

Table 7 presents for the individual groups the control parameter β' as well as the thermodynamical variables S_N, U_N/κ and F_N/Q for the two interaction conditions. The sample averages were reported in Table 8. For the 'eyes closed' condition the control parameter β' was close to zero ($M = 0.1$). In contrast, for the 'intentional' interaction condition β' was much larger ($M = 2.7$). The 'disorder' measure, entropy S_N, was relatively high for the 'eyes closed' condition ($M = 9.19$), whereas the 'disorder' was relatively low for the 'intentional' condition ($M = 7.6$). The internal energy was close to zero for the 'eyes closed' condition ($M = -0.01$) and dropped dramatically for the 'intentional' condition ($M = -2.9$). Likewise, the free energy was lower for the 'intentional' condition ($M = -10.6$) than for the 'eyes closed' condition ($M = -9.19$). Paired sample t tests showed that the difference between the control parameters was statistically significant ($t(5) = 5.190, p < 0.01$). Likewise, the effect of the interaction type on the 'disorder' measure S_N was significant ($t(5) = 3.691$, $p < 0.05$). The interaction type had a significant effect on the internal energy ($t(5) = 3.577, p < 0.05$) and the free energy ($t(5) = 3.447$, $p < 0.05$).

Table 7. Control parameter and thermodynamical variables estimated for individual groups under two interaction conditions. In order to improve the presentation, the first and forth columns of Table 5 are presented here again.

Group	'Eyes closed'					'Intentional' Av.(1+2)				
	$\langle y \rangle_T$	β'	S_N	U_N/κ	F_N/Q	$\langle y \rangle_T$	β'	S_N	U_N/κ	F_N/Q
1	0.21	0.10	9.19	0.00	−9.19	0.37	1.40	8.87	−0.59	−9.45
2	0.24	0.45	9.17	−0.05	−9.21	0.36	1.35	8.89	−0.54	−9.43
3	0.20	0.00	9.19	0.00	−9.19	0.67	3.30	7.09	−3.91	−11.00
4	0.20	0.00	9.19	0.00	−9.19	0.61	2.85	7.57	−2.94	−10.50
5	0.20	0.00	9.19	0.00	−9.19	0.74	3.95	6.48	−5.36	−11.83
6	0.20	0.00	9.19	0.00	−9.19	0.69	3.45	6.94	−4.25	−11.19

Table 8. Sample means of the scores reported in Table 7

Condition	$M_{\bar{y}}$	M_{β}	M_{S_N}	$M_{U_N/\kappa}$	$M_{F_N/Q}$
Eyes closed	0.21 (0.02)	0.1 (0.2)	9.19 (0.01)	−0.01 (0.02)	−9.19 (0.01)
Intentional	0.57 (0.17)	2.7 (1.1)	7.6 (1.0)	−2.9 (2.0)	−10.6 (1.0)

4.5. Discussion

While in Frank and Richardson [27] group coordination has been quantified on the basis of a descriptive measure, the squared cluster amplitude y, we demonstrated that a model-based analysis based on a thermo-statistical interpretation of the Kuramoto model can be conducted. In Frank and Richardson [27], it was shown that in the 'eyes closed' condition the squared cluster amplitude y was not statistically significant different from the test value expected under the hypothesis of completely de-synchronized participant group. In contrast, the squared cluster amplitude y observed under the 'intentional' condition was significantly different from the aforementioned test value. Consistent with this finding, we found that the control parameter β' was significantly higher in the 'intentional' condition and the measure for the behavioral disorder S_N among the group members was significantly lower in the 'intentional' condition. Interestingly, we also found a statistically significant decay of the free energy F_N in response to the intended interaction between the group members. This is consistent with the Landau free energy which not only predicts that in the stationary case free energy is minimized but also predicts that the minimal free energy decays further away from the non-equilibrium phase transition point (see Sec. 1. and Figure 2, right).

5. Coda

A general theoretical framework to stochastic motor coordination was presented that is based on the notion of thermodynamical forces $\mathbf{X}_{\mathrm{th}}$ that induce thermodynamical flows $\mathbf{J}_{\mathrm{th}}$. It was shown that this concept allows us to establish a link between thermodynamical quantities such as energy and entropy and stochastic, dynamical descriptions of human motor coordination. Essential to this approach was the fact that during transient periods free energy decays due to subprocesses that are internal to the motor control system and can only increase entropy or leave it invariant. This implies that the proposed approach exhibits an in-built time arrow. During transient periods, processes evolve in one direction not in the reverse. This allows to account for stability of motor performance (recall that in this context we reported in the study of uni-manual oscillatory limb movements about two stability parameters: γ and λ^*). Furthermore, the decay of free energy during transient periods implies the existence of a free energy principle. This, in turn, is an appealing property of the theoretical framework because the well-established Landau theory of equilibrium phase transitions exhibits the same kind of principle.

Moreover, we showed by means of our two explicit examples that minimal free energy measures estimated from experimental data decay when an appropriately defined control parameter is scaled further and further away from the bifurcation point that describes the onset of the coordinated motor activity. This observation is consistent with the aforementioned Landau theory. The generality of this observation is open for debate. For example, in the study on uni-manual rhythmic wrist movements the analysis of the internal energy suggested that the pendulum resonance frequency may play a crucial role how thermodynamical variables scale with the control parameter. Future research may be devoted to determine whether or not the minimal free energy exhibits a non-monotonic behavior close to such a resonance.

The approach discussed in the previous sections accounts for the fact that during transients internal processes may result in an increase of entropy. In the stationary case, forces and fluxes equal zero such that the entropy production approaches zero as well. This might be unrealistic in some applications. For example, rhythmic coordinated motor activity such as walking results in entropy production even when the walker has settled down in a stationary walking pattern. In order to account for entropy production involved in an on-going coordinated activity, energy pumping processes must be taken into consideration [21].

As reviewed in the introduction, motor coordination has been regarded as a process of self-organization and an instance of a dissipative structure emerging in a bifurcation. Accordingly, the pattern of coordinated activity exhibits a spatio-temporal structure. We may say it exhibits some degree of order. The notion of an emerging pattern exhibiting order has lead to the speculation that entropy (as a measure of disorder) must decay when the coordination emerges. In our two examples, it seems as if we arrived at inconclusive results. In the study on uni-manual rhythmic activity, entropy increased when pacing frequency was increased (see the measure S_H in Table 4). In contrast, entropy decreased in the study on group coordination when group members intentionally coordinated their rocking movements (see Table 8). However, the problem with these seemingly inconclusive results can be resolved. As pointed out by Haken, entropy does not necessarily decay at the onset of a self-organized behavior [34]. The reason for this is that the entropy measure S is scale dependent. A coordinated, ordered behavior may exhibit a large scale that induces an increases of S. In general, it is recommended to examine the behavior of the entropy separately for each system under consideration in order to check whether it is plausible to assume that the emergence of order can be captured and quantified in terms of a decrease in entropy S.

References

[1] J. A. Acebron, L. L. Bonilla, C. J. P. Vicente, F. Ritort, and R. Spigler. The Kuramoto model: a simple paradigm for synchronization phenomena. *Rev. Mod. Phys.*, 77:137–185, 2005.

[2] P. J. Beek, C. E. Peper, and D. F. Stegeman. Dynamical models of movement coordination. *Hum. Movement Sci.*, 14:573–608, 1995.

[3] P. J. Beek, W. E. I. Rikkert, and P. C. W. van Wieringen. Limit cycle properties of rhythmic forearm movements. *J. Exp. Psychol. - Hum. Percept. Perform.*, 22:1077–1093, 1996.

[4] P. J. Beek, R. C. Schmidt, A. W. Morris, M.-Y. Sim, and M. T. Turvey. Linear and nonlinear stiffness and friction in biological rhythmic movements. *Biol. Cybern.*, 73:499–507, 1995.

[5] N. A. Bernstein. *The coordination and regulation of movements*. Pergamon, Oxford, 1967.

[6] J. Buck and E. Buck. Mechanism of rhythmic synchronous flashing of fireflies. *Science*, 159:1319–1327, 1968.

[7] J. Buck and E. Buck. Synchronous fireflies. *Sci. American*, 234(5):74–85, 1976.

[8] S. Chandrasekhar. *Liquid crystals*. Cambridge University Press, Cambridge, 1977.

[9] A. Compte and D. Jou. Non-equilibrium thermodynamics and anomalous diffusion. *J. Phys. A: Math. Gen.*, 29:4321–4329, 1996.

[10] S. R. de Groot and P. Mazur. *Non-equilibrium thermodynamics*. North-Holland Publ. Company, Amsterdam, 1962.

[11] D. G. Dotov and T. D. Frank. From the W-method to the canonical-dissipative method for studying uni-manual rhythmic behavior. *Motor Control*, 15:550–567, 2011.

[12] W. Ebeling and I. M. Sokolov. *Statistical thermodynamics and stochastic theory of nonequilibrium systems*. World Scientific, Singapore, 2004.

[13] B. Eckhardt, E. Ott, S. H. Strogatz, D. M. Abrams, and A. McRobie. Modeling walker synchronization on the Millenium bridge. *Phys. Rev. E*, 75:021110, 2007.

[14] U. Erdmann, W. Ebeling, L. Schimansky-Geier, and F. Schweitzer. Brownian particles far from equilibrium. *Eur. Phys. J. B*, 15:105–113, 2000.

[15] T. D. Frank. Generalized Fokker-Planck equations derived from generalized linear nonequilibrium thermodynamics. *Physica A*, 310:397–412, 2002.

[16] T. D. Frank. Generalized multivariate Fokker-Planck equations derived from kinetic transport theory and linear nonequilibrium thermodynamics. *Phys. Lett. A*, 305:150–159, 2002.

[17] T. D. Frank. On the boundedness of free energy functionals. *Nonlin. Phenom. Complex Syst.*, 6(3):696–704, 2003.

[18] T. D. Frank. *Nonlinear Fokker-Planck equations: Fundamentals and applications*. Springer, Berlin, 2005.

[19] T. D. Frank. Linear and nonlinear Fokker-Planck equations. In R. A. Meyers, editor, *Encyclopedia of Complexity and Systems Science*, volume 5, pages 5239–5265, Berlin, 2009. Springer.

[20] T. D. Frank. On a moment-based data analysis method for canonical-dissipative oscillator systems. *Fluctuation and Noise Letters*, 9:69–87, 2010.

[21] T. D. Frank. Pumping and entropy production in non-equilibrium drift-diffusion systems: a canonical-dissipative approach. *European Journal of Scientific Research*, 46:136–146, 2010.

[22] T. D. Frank. Strongly nonlinear stochastic processes in physics and the life sciences. *ISRN Mathematical Physics*, 2013:article 149169, 2013.

[23] T. D. Frank, D. G. Dotov, and M. T. Turvey. A canonical-dissipative approach to control and coordination. In F. Danion and M. Latash, editors, *Motor control: theories, experiments, and applications*, pages 50–71, New York, 2010. Oxford Press.

[24] T. D. Frank, R. Friedrich, and P. J. Beek. Stochastic order parameter equation of isometric force production revealed by drift-diffusion estimates. *Phys. Rev. E*, 74:051905, 2006.

[25] T. D. Frank, S. Kim, and D. G. Dotov. Moment method with a MATLAB script for the parameter estimation of canonical-dissipative nonequilibrium energy distributions in the life sciences, submitted.

[26] T. D. Frank and S. Mongkolsakulvong. Parametric solution method for self-consistency equations and order parameter equations derived from nonlinear fokker-planck equations. *Physica D*, 238:1186–1196, 2009.

[27] T. D. Frank and M. J. Richardson. On a test statistic for the Kuramoto order parameter of synchronization with an illustration for group synchronization during rocking chairs. *Physica D*, 239:2084–2092, 2010.

[28] C. W. Gardiner. *Handbook of stochastic methods*. Springer, Berlin, 2 edition, 1997.

[29] P. Glansdorff and I. Prigogine. *Thermodynamic theory of structure, stability, and fluctuations*. John Wiley and Sons, New York, 1971.

[30] R. Graham. Onset of cooperative behavior in nonequilibrium steady states. In G. Nicolis, G. Dewel, and J. W. Turner, editors, *Order and fluctuations in equilibrium and nonequilibrium statistical mechanics*, pages 235–288, New York, 1981. John Wiley and Sons.

[31] R. Graham and T. Tel. Existence of a potential for dissipative dynamical systems. *Phys. Rev. Lett.*, 52:9–12, 1984.

[32] H. Haken. Distribution function for classical and quantum systems far from thermal equilibrium. *Z. Physik*, 263:267–282, 1973.

[33] H. Haken. *Principles of brain functioning*. Springer, Berlin, 1996.

[34] H. Haken. *Information and self-organization*. Springer, Berlin, 2 edition, 2000.

[35] K. Huang. *Statistical mechanics*. John Wiley and Sons, New York, 1963.

[36] V. K. Jirsa and J. A. S. Kelso. *Coordination dynamics: Issues and trends*. Springer, Berlin, 2004.

[37] E. E. Kadar, R. C. Schmidt, and M. T. Turvey. xxx. *Biological Cybernetics*, 68:421–430, 1993.

[38] B. A. Kay, J. A. S. Kelso, E. L. Saltzman, and G. Schöner. The space-time behavior of single and bimanual movements: Data and model. *J. Exp. Psychol. - Hum. Percept. Perform.*, 13:178–192, 1987.

[39] J. A. S. Kelso. *Dynamic patterns - The self-organization of brain and behavior*. MIT Press, Cambridge, 1995.

[40] D. Kondepudi and I. Prigogine. *Modern thermodynamics*. John Wiley and Sons, New York, 1998.

[41] R. Kubo. *Statistical Mechanics*. North-Holland Publ. Company, Amsterdam, 1967.

[42] P. N. Kugler and M. T. Turvey. *Information, natural law, and the self-assembly of rhythmic movement*. Erlbaum, Hillsdale, New Jersey, 1987.

[43] P. N. Kugler, M. T. Turvey, R. C. Schmidt, and L. D. Rosenblum. Investigating a nonconservative invariant of motion in coordinated rhythmic movements. *Ecological Psychology*, 2:151–189, 1990.

[44] Y. Kuramoto. *Chemical oscillations, waves, and turbulence*. Springer, Berlin, 1984.

[45] L. D. Landau and E. M. Lifshitz. *Statistical physics*. Pergramon Press, London, 1958.

[46] K. L. Marsh, M. J. Richardson, and R. C. Schmidt. Social connection through joint action and interpersonal coordination. *Topics in Cog. Sci.*, 1:320–339, 2009.

[47] L. K. Miles, L. K. Nind, and C. N. Macrae. The rhythm of rapport: interpresonal synchrony and social perception. *J. Exp. Social Psychology*, 45:585–589, 2009.

[48] S. Mongkolsakulvong and T. D. Frank. Canonical-dissipative limit cycle oscillators with a short-range interaction in phase space. *Condensed Matter Physics*, 13:13001, 2010.

[49] Z. Neda, E. Ravasz, Y. Brechet, T. Vicsek, and A. L. Barabasi. The sound of many hands clapping. *Nature*, 403:849–850, 2000.

[50] Z. Neda, E. Ravasz, T. Vicsek, Y. Brechet, and A. L. Barabasi. Physics of the rhythmic applause. *Phys. Rev. E*, 61:6987–6992, 2000.

[51] A. Pikovsky, M. Rosenblum, and J. Kurths. *Synchronization: a universal concept in nonlinear sciences*. Cambridge University Press, Cambridge, 2001.

[52] M. Plischke and B. Bergersen. *Equilibrium statistical physics*. World Scientific, Singapor, 1994.

[53] F. Reif. *Fundamentals of statistical and thermal physics*. McGraw-Hill Book Company, New York, 1965.

[54] M. J. Richardson, R. L. Garcia, T. D. Frank, M. Gergor, and K. L. Marsh. Measuring group synchrony: a cluster-phase method for analyzing multivariate movement timeseries. *Frontiers in Physiology*, 3:405, 2012.

[55] M. J. Richardson, K. L. Marsh, R. W. Isenhower, J. R. L. Goodman, and R.C. Schmidt. Rocking together: dynamics of intentional and unintentional interpresonal coordination. *Hum. Movement Sci.*, 26:867–891, 2007.

[56] M. J. Richardson, K. L. Marsh, and R. C. Schmidt. Effects of visual and verbal information on unintentional interpresonal coordination. *J. Exp. Psychol. - Hum. Percept. Perform.*, 31:62–79, 2005.

[57] H. Risken. *The Fokker-Planck equation — Methods of solution and applications*. Springer, Berlin, 1989.

[58] R. C. Schmidt, C. Carello, and M. T. Turvey. Phase transitions and critical fluctuations in the visual coordination of rhythmic movements between people. *J. Exp. Psychol. - Hum. Percept. Perform.*, 53:247–257, 1990.

[59] R. C. Schmidt and M. J. Richardson. Dynamics of interpersonal coordination. In A. Fuchs and V. Jirsa, editors, *Coordination: neural, behavioral and social dynamics*, pages 281–308, Heidelberg, 2008. Springer.

[60] R. C. Schmidt and M. T. Turvey. Phase entrainment dynamics of visually coupled rhythmic movements. *Biol. Cybern.*, 70:369–376, 1994.

[61] K. Shockley, M. J. Richardson, and R. Dale. Conversation and coordinative structures. *Topics in Cog. Sci.*, 1:305–319, 2009.

[62] G. Strobl. *Condensed matter physics — Crystals, liquids, liquid crystals, and polymers*. Springer, Berlin, 2004.

[63] S. H. Strogatz. From Kuramoto to Crawford: exploring the onset of synchronization in populations of coupled oscillators. *Physica D*, 143:1–20, 2000.

[64] S. H. Strogatz, D.M. Abrams, A. McRobie, B. Eckhardt, and E. Ott. Crowd synchrony on the Millennium Bridge. *Nature*, 438:43–44, 2005.

[65] M. T. Turvey. Coordination. *Am. Psychol.*, 45:938–953, 1990.

[66] C. van den Berg, P. J. Beek, R. C. Wagenaar, and P. C. W. van Wieringen. Coordination disorders in patients with Parkinson's disease: a study of paced rhythmic forearm movements. *Exp. Brain Res.*, 134:174–186, 2000.

[67] G. H. Wannier. *Statistical physics*. Dover Publications, New York, 1966.

[68] A. T. Winfree. *The geometry of biological time*. Springer, Berlin, 2 edition, 2001.

Index

A

B

C

D

E

F

G

H

I

J

K

L

M

N

O

P

T

U

V

W

Y